Stewardship & Conservation in Agriculture

The Small Farm Reality

THE AUTHORS

Dr. Moumita Dey Gupta received M.Sc.(Ag.) in Agricultural Extension (2014), Ph.D. in Agricultural Extension (2018) under the guidance of Prof S K Acharya in the Dept. of Agricultural Extension, from Bidhan Chandra Krishi Viswavidyalaya and Post Graduate Diploma in Journalism and Mass Communication (2017) from Netaji Subhas Open University. She was a recipient of the DST INSPIRE Fellowship for securing the highest overall grade point in M.Sc. (Ag.) in Agricultural Extension (2015). She has qualified ICAR and UGC NET in the year 2016. She has already authored 2 more books. She has also contributed some book chapters and a good count of research papers in national and international Journals. Presently, she is working as Subject Matter Specialist (Agricultural Extension) at WBCADC, KVK Bankura since 2018 and is disseminating new technologies through leaflet, folder, newspaper, and radio broadcast in vernacular for the farming community of the state with a great zeal.

Prof. Sankar Kr Acharya, former Head, Dept. of Agril Extension and Director, Extension Education, Bidhan Chandra Krishi Viswavidyalaya, Mohanpur, WB, born on 6th October, 1960, started his career as Assistant Professor at BCKV in 1988 and has been in teaching, research and extension over 30 years. Research Publication: 217 in National and International Journals. Book Publication: 99 books authored/ co-authored Membership of Distinction: Chancellor Nominee on selection committee for professors' positions; Member of Agriculture Commission; Marketing and Extension Sub Committee, Govt. of West Bengal; Expert member WWF projects in BTR (Buxa Tiger Reserve Project); Expert member; DFID project on Primary Education (DPEP); REIA, Visited Italy, France, Germany, China, Sri Lanka and Bangladesh. Delivered 18 Key note addresses in International Congress, Chaired 19 Sessions in International conferences, Editors of a score of national and international journals. He has so far been awarded with 13 best paper awards in National and International Conferences Organizing Secretary of two International Conferences, He has also been honoured to be selected as the Convener of Panel (PE-32) entitled The hunger , poverty and silence of IUAES, University of Manchester, UK, 2013. He is the fellow, ISEE, IARI, New Delhi.

Dr. Amit Baran Sharangi is a Professor of BCKV (Agricultural University) attached with teaching, research and extension for about 23 years. He completed his Post-Doctorate from the University of Melbourne, Australia. He is specially trained from IISR, IIHR & ARO (Tel Aviv, Israel); experienced in herbs, spices and medicinal-aromatic plants. He worked at laboratories of Australia, USA and UK for continuing research on those crops as well as teaching Post Graduate students in the universities abroad. Prof Sharangi has several international awards including FULBRIGHT (USA), INSA-RSE Visiting Scientist (UK), Endeavour Award (Australia), Marquis Who's Who (USA), Young Achievers Award (SADHNA), Higher Education Leadership Award, Outstanding Scientist Award, Bharat Ratna Mother Teresa Gold Medal Award, etc. He is Fellow of eight prestigious academies/societies and members of many international bodies having academic repute. He has published about 85 peer-reviewed papers, 25 books and numerous book chapters within his subject area. He is also associated with 50 journals worldwide as editor-in-chief, associate editor, regional & technical editor, editorial board member and active reviewer. He has administrative experience for four years as HOD, chaired several sessions of national and international conferences, run six projects so far as PI and Co-PI, developed as well as restructured PG courses as BSMA committee member (DG, ICAR, Govt of India) at national level.

Stewardship & Conservation in Agriculture

The Small Farm Reality

Moumita Dey Gupta, *Ph.D*
Sankar Kr Acharya, *Ph.D*
A B Sharangi, *Ph.D*

2022

Scholars World

A Division of

Astral International Pvt. Ltd.

New Delhi – 110 002

ISBN: 9789354615412

Published by : **Scholars World**
A Division of
Astral International Pvt. Ltd.
– ISO 9001:2015 Certified Company –
4736/23, Ansari Road, Darya Ganj,
New Delhi-110 002
Ph. 011-4354 9197, 8130496929
E-mail: info@astralint.com
Website: www.astralint.com

Professor K V Peter

Ph.D., FNAAS, FNASc FNABS. FISVeg, FISGen., FHSI, H.FCHAI FISNS

Former Vice-Chancellor Kerala Agricultural University and Director ICAR IISR Calicut

FOREWORD

"India lives in farmers huts"-Mahatma Gandhi. "Agriculture is the most beautiful, most useful and most noble employment of men"-George Washington. "If conservation of natural resources goes wrong, nothing else will go right"-M. S. Swaminathan. The above quotes form the basics of the present book "Stewardship and Conservation in Agriculture: The Small farm Reality"

The good earth has her own and the sweetest music of ecological resilience and rhythm, as if, the symphonies of Beethoven or Mozart are on their best of melodies. Agriculture, once upon a time, has been the pristine child of nature with all resonance of traditional knowledge, praxis and natural science. The explosion of population and industrialization, consumerism and urbanization have driven us to a compulsion of making ecosystem denuded, depleted and polluted with all kinds of disruptions inflicted into soil, water and biodiversity. By 2050, agriculture has to feed 9.5 billion population, billions of livestock's and birds from exponentially squeezing land: man ratio. Every year we are losing 10-15 tonnes of topsoil per ha. of land in India to contribute fathomless volume of erosion of fertile soils due to invasion of farm machineries with all the wrong notes and dicta. Some of wrong doings executed in our 'modernizing agriculture' have left behind us an array of learning experiences. The most neglected aspects and issues are the negative energy metabolism in soil ecosystem, disruptions in carbon, nitrogen and water cycles, absence of an ensured and happy return to farmers of marginal economy, and, above all, the perturbations of ecological resilience. I strongly believe the book, Stewardship and Conservation in Agriculture: The Small farm Reality edited by Prof S K Acharya, Moumita De Gupta and Prof. A B Sarangi is a well endeavoured intellectual exercise with unique hard evidences to make it a worth reading experiences for scholars across the disciplines. Congratulations to all the contributing authors for such a splendid book creation. The authors of chapters are working scientists in the field. The publisher Astral International New Delhi is a well established publishing house and they have printed out an outstanding book. I wish readers in ICAR Institutes, Central and State Agricultural Universities will be much bend-Med.

K V Peter

PREFACE

The global agriculture since past few decades has been seriously suffering from yield decline, resource erosion and economic impoverishment. A paradigm shift is essentially imminent to save the resource base of agriculture in order to save the existence of small and marginal farmers, especially from South East Asian countries. The present book concerns the serious decadence and impoverishment as evinced in the increasing rural poverty due to soil erosion, water contamination and scarcity, undulating market prices and brunt of climate change as shaking the future of agriculture and food security in India. While the national economy is focusing on escalation of food production by triggering technology-driven and ICT supported agricultural modernization, the non-remunerative agricultural occupation as persuaded by small and marginal farmers have been thrown into a heartless conflict between eroding resource bases and escalating input prices to make agriculture a profitable choice for young minds in this part of the planet. The stubbles are burnt to ashes to deprive the soil from its right to get back the crop residues through carbon sequestration. The energy prodigal nature can only be attributed to the reckless fragmentation of 143 million hectares of arable land being mentored by 121 million of farm holders with an average of 1.40 hectares of holding size. Is it that an approach of conservation and resource stewardship cannot be an acceptable proposition for small land owners with further denudations and fragmentations! Does the future farming for poorer farmers are going to be a valley of conflict between ecology and economy or, saving water versus producing more grains by depleting more ground water with a congenital threat of climate variance. So, the global audiences are keen to hear the saga of ecological stewardship by small farmers not withstanding a sharp resistance from hunger and poverty remains an irresistible concern. This book, Stewardship and Conservation in Agriculture: The Small farm Reality, offers a unique cultivation of mind as to how we can go for stewardship and conservation policy and intervention for small farm reality. It has gone far beyond the 'romanticized' farm stewardship, for bigger holding sizes, say more than one hectare of land, yet to take account the feasibility of stewardship operations for tiny holdings, 1-1.5 hectares as predominant in most of the Asian countries based on hard evidences. While American and European farms are mostly belonging to consolidated and big size holdings, Asian part of global agriculture has been coercive in the form of soil erosion, water contamination and dwindling farm income, which are mostly due to marginalization and impoverishment of both income and livelihood from farm. This is really a difficult task, of course is of huge scholastic interest, for making the farmers understand that through proper stewardship and conservation interventions the invaluable resources like land, water and biodiversity need to be protected and put to constant regenerating practices. So, the book includes a unique periphery of stewardship management wherein the issues of farm income, livelihood and ecological restorations are in seamless contradictions and conflicts. And, that is why the book certainly merits unique empirical as well as pedagogical texts for the scholars and policy makers.

Kalyani

S K Acharya

CONTENTS

CHAPTER 1

INTRODUCTION

If any economic sector in the world has certainly been poised for a dramatic change from its classical tradition to highly modernizing pursuits, Indian agriculture is obviously the zenith of an example. For past two decades, a clear swing for Indian agriculture has been in favor of creating and venturing agri-preneurships over augmenting agriculturalproductivity only. While the issues of food security for approximately 1300 million of her teeming populace, India has also to ensure the livelihood security for 450 million of holding and farm families, (The Hindustan Times , 2014)

Indian Agriculture has got three distinct phases of structural and functional evolution and these are- Inductive phase of agriculture (1950-1970), Stimulative phaseof agriculture (1970-1990) and Simulative phase of agriculture (1990 onwards). For inductive phase, we had a clear dictum for augmenting the productivity with the highest intensity of inputs so that less time has been required to reach the target of food production at the earliest. Hunger and poverty for this period (1950-1970) had exerted a looming effect and posed threats to our sociopolitical as well as economic stability. Atthe end of this phase, we had earned the swashbuckling and socially rippling "Green - Revolution". A quantum jump, from a 55 million tonnes of food grain production to 242.73 million tonnes of food grain (Agricultural Statistics at a Glance, 2014), has certainly been an agricultural big-bang. We happily declared and proclaimed the green revolution and the same approaches followed in post green revolution phases till 1990s. Now we have started thinking that if one side of the coin is green revolution, the other side must be a grey zone.

The much proclaimed green revolution has to confront a plethora of criticisms. The indiscriminate use of fertilizer, pesticides, plant hormones etc. has certainly contributed to produce bigger volume of food; the other side of the episode reveals thatat least 45 percent of ground water has been contaminated and almost an equivalent percentage of land is under the coercive process of salinization as well as desertification. Once blessed with more than 2 lakh (2, 00000) rice varieties, the myopic approaches in modern agriculture driven by high yielding varieties and hybrid vigour in rice bio- genomics followed in the green revolution has led to serious crop biodiversity erosionand approximately 416 Indian traditional rice varieties, are on the verge of extinction from farmers' fields (Mohammed Ashraf, 2017).

Soil, water and biodiversity are the main pillars for agriculture to grow and survive. But as an aftermath of green revolution, associated with unabated urbanization, soil erosion has gone to its worst stage. 5334 million tons of soil is eroded per ha of land every year, which itself is a serious threat to agricultural sustainability (The Hindu, 2010). The problem has further been worsened by unabated population pressure. The availability of land per family has been downsized to 1.15 ha, which again has declined further to 0.3 ha for West Bengal (Agriculture census 2010-11 and West Bengal development report 2010).The brunt of climate change and global warming has opened up new phases of threats and challenges. The monsoon behavior has changed with a 10- 12 percent swing in favour of post monsoon precipitation; the other threat is certainly theincrease of night temperature. With every 1°C change in night temperature will contribute to 10 (Peng et al, 2004) per cent yield loss for rice and 6.0 ± 2.9 per cent for wheat (Zhao et al, 2017). When climate is changing, the environmental behavior is becoming more unpredictable. Some of the disease pest is offering a devastating come back in a response to increase of temperature and change of rainfall pattern.

Another alarming issue for present Indian agriculture is the farmers' income and livelihood. While we have set a national goal for doubling of farmer's income by 2022, the question of revamping institutional structure and proper policy formulations will certainly be more strident. In India, all costs are certain for farmers and every income for them is uncertain. More than 3 lakh farmers have committed suicide to generate a colossal impact and apocalyptic effect on the very existence of farmers in some parts of India (NCRB, 2016). When institutional failures reached a dire stage, the issues of sustainability go further bleak and bizarre.

With these perspectives, the question of conservation and stewardship is both conspicuous and pertinent. Without conserving soil, the fertility and productivity cannot be achieved. At the same time, conservation of ground water is essential to ensure and defend irrigation water to at least 88.63 million hectares of land under being managed under irrigated agro ecosystem in India (Directorate of economics and statistics, Ministry of Agriculture.). For rain fed agro ecosystem, watershed management along with rainwater harvesting shall go in a congenital manner. Now, the two aspects of conservation and stewardship need to be operationally discriminated although they have some agenda to share. While conservation approaches, being classical in nature, aimingat maintaining the count of biota of physical resources, the stewardship function extrapolates conservation in redressing anthropogenic needs and its peripherals. In India, even if conservation is known to certain sector of farming Diasporas, the concept and role of stewardship function remain bit alien and stranger. What is there if incase forest resources are conserved unless it serves human need!

So, attuning and upgrading conservation practices with stewardship function are adaptable and socializable functions to attain a sustainable growth and development. This is the process where in forest is protected from its biodiversity depletion and at the same time the part of the forest is opened up for ecotourism. Thus, conservation elementsare beautifully extended into stewardship functions through the generation of both income and livelihood. The changing paradigm from conservation to stewardship needs areciprocal change in the text and dicta of community participation, local governance, agricultural policy and political interventions.

Stewardship function does not entertain a mundane as well as insipid conservation approaches; it really welcomes ecological resilience by incubating and internalizing ecological deliverables to support both human and non-human life forms. Ecological functions are basically the productive performance of ecological deliverables for the surrounding social- ecosystem to make it an agile, dynamic and ever sproutinglife support system. With this background the present study has been designed and directed with the following rationale:

i. We need to have a paradigm shift from conservation to stewardship approaches so that the natural resource conserved can go in a resilient manner to support income, livelihood and social progress in interactive dispositions.

ii. Our farmers are mostly input concerned and less energy conscious and the least concerned on stewardship aspect.

iii. The soil, water and biodiversity are the prime movers to usher the transformation process of conservation approaches into stewardship functions and for which we need to generate baselines as well as preparedness of the subject farmers as to how and where they stand for stewardship functions.

iv. The stewardship functions for this empirical study can be estimated through different farm operation, management, indigenous knowledge, praxis, community innovation and social networking.

v. Whether the issues can be redressed in terms o non-cognitive stewardship with cognitive stewardship; functional stewardship with both non cognitive and cognitive, and ecological deliverables with stewardship functions and thereafter a disposition of dynamic interplay between ecology and sociology.

Hence the objectives are as follows:

i. To conduct a baseline survey for the research locale in order to identify appropriate respondents for the study.

ii. To identify appropriate variables suitable for measuring the perception of therespondents towards conventional conservation practices in the study area.

iii. To estimate the inter- and intra level relationship among and between the set ofvariables to finally elucidate the perception of conservation practices.

iv. To assess the perception of the respondents on conventional conservation practices as well as possibility of stewardship.

v. To formulate extension intervention based on results.

2

CHAPTER

CITATION AND RELATED LITERATURE

This chapter reviews the research work done in the fields related to the objectives of the study. In any scientific investigation or research a comprehensive review of relevant literature is very essential. Its main function apart from determining the work done before concerning the problem area i.e. area of investigation, it provides an insight into the methods and procedures and create a basis for interpretation of findings. As direct references on all the item are not in abundance, certain specific references along with some indirect references have been incorporated this chapter for the purpose of meaningful use.

Sl. No.	Author	Publishing Year	Title	Source	Abstract
1.	Ditsch, D C; Trimble, R L; Wade,J M	1988	Tillage selection:soil stewardship versus financial survival.	Special-Bulletin, Mississippi Agricultural and Forestry Experiment-Station, 1988; (88-1): 5-8	Crop production budgets are produced for conventional, reduced and no-tillage systems, and after change to reduced and no-till systems. It is concluded that although conservation tillage offers long-term advantages, in the short term yields may be reduced, forcing crop producers into financial hardship and hence the system is not used by many Kentucky farmers.
2.	Washington, D.C., USA: National AcademyPress.	1988	Understanding agriculture: newdirections for education.	Understanding-agriculture:new directions for education. 1988; ix + 68pp.	The committee's findings point to two basic challenges: first, agricultural education must become more than vocational agriculture. Second, major revisions are needed within vocational agriculture. In working toward both goals, educators should borrow from the best current programmes, while creating new ways to deliver to more students' educational opportunities in the agricultural sciences, agribusiness, nutrition, and land resource stewardship.

Sl. No.	Author	Publishing Year	Title	Source	Abstract
3.	Brklacich, M J	1989	A framework for evaluating the sustainability of food production systems in a changing environment.	Dissertation Abstracts International A, Humanities and Social Sciences. 1989; 50(6): p.1764	A sustainable food production system is defined as an agri-food system which simultaneously maintains environmental quality, provides acceptable economic and social rewards to agricultural producers, and ensures an adequate and accessible food supply. A three-dimensional framework for evaluating the sustainability of food production systems is developed comprising the resource stewardship dimension, the producer- community dimension, and the food sufficiency dimension. It was found that reductions in crop yields owing to soil erosion and climatic warming would reduce farm income potential by 26% and regional production capacity would fall short of 1981 production levels by 14%.
4.	Grossi, R E	1993	A green evolution: retooling agricultural policy for greater sustainability.	Journal of Soil and Water Conservation. 1993; 48(4): 285-288	The emergence of a sustainable agriculture will require considerable changes. A deliberate and gradual shift in policy priorities, what might be called a 'green evolution' of agriculture, toward principles of resource stewardship and marketplace economics, which must be recognized as not mutually exclusive goals, is necessary..
5.	Fox, G	1994	The structure ofownership and prosperity	Canadian Journal of Agricultural Economics. 1994; 42(4): 511-524	The structure of property ownership does have important implications for resource stewardship and particularly, the institution of 'several property' plays an essential role in the coordination of human interaction. It is concluded, however, that further development is necessary. Clarification of the basisand contents of rights is also required.
6.	L M; Liggett, N; Havens, D; Anderson, W C; Stark, L K	1994	Agricultural stewardship: building partnerships through a focus group process. An overview of the WSU SkagitCounty	Extension Bulletin Cooperative Extension, College of Agriculture and Home Economics, Washington State Universi ty. 1994;	Partnerships between farmers, consumers, agricultural industry representatives, environmentalists, government and university personnel in the form of focus groups can take advantage of creative group energy to jointly identify needs and set priorities for agricultural research and education. It was concluded that the focus group approach offers some

Sl. No.	Author	Publishing Year	Title	Source	Abstract
			Water Quality and Nitrate Leaching Focus Group... a model for building partnerships between diverseinterest groups.	(EB1788): 19 Pullman, Washington, USA	Novel possibilities to the land grant university system for re-thinking the way in which it sets and carries out its agenda. There is a need to establish problem solving teams that include a wide range of interests, some of which may have been excluded in the past.
7.	Colman, D	1994	Ethics and externalities: agricultural stewardship andother behaviour: presidential address.	Journal of Agricultural Economics. 1994; 45(3): 299-311	Neoclassical microeconomic theory of the firm and household rests on the foundations of profit maximizing and utility maximizing behaviour respectively. After exploring the issue of whether market efficiency can be improved when economic agents' behaviour is motivated by ethical objectives such as fairness and altruism, agricultural stewardship came up as an ethically motivated form of behaviour, which is profit sacrificing, but which improves economic efficiency by reducing environmental externalities.
8.	Regan, J B	1995	Natural resourcestewardship efficacy of Treflan and Sonalan in conservation tillage.	Down to Earth Midland. 1995; 50(2): 6-12	The efficacy of Treflan in no till wheat, results indicated that Treflan 10 gave adequate control of S. indica with high dosage and/or appropriate timing, and gave better control when incorporated than Treflan MTF, whilst residues were maintained at acceptable levels. Best residue conservation occurred when cultivation was done by disc, combination tool, rotary hoe or Haybuster under cutter. The subsequent wheat crop was unaffected by any of the treatments.
9.	Ticknor, W D	1996	Arkansas forestry: maintaining the momentum of excellence.	Special Report. Agricultural Experiment Station, Division of Agriculture, University of Arkansas. 1996; (175): v +10 pp. Fayetteville,Arkansas, USA:	The first Arkansas Forest Resources Distinguished Lecture deals with excellence in resource stewardship, land ethics, ecosystem management, systems thought and several other issues relevant to modern forest management in the particular context of Arkansas.
10.	El Baltagy, A	1996	West Asia and North Africa: a regional vision.	West Asia and North Africa: a regional vision. 1996; 16 pp, Aleppo, Syria:International Center for Agricultural Research in the Dry Areas.	Poverty, agricultural employment and migration; food consumption and production; natural resource stewardship; and global economic integration and food security are of great concern and in which we should focus to meet the challenge ofthe current levels of development, to halt the escalating food gap, and tackle rural poverty and natural resource managementproblems.

Sl. No.	Author	Publishing Year	Title	Source	Abstract
11.	Hoogheem, T J	1997	Global product stewardship water quality protection.	Brighton Crop Protection Conference: Weeds Proceedings of an international conference, Brighton, UK, 17 20 November 1997. 1997; (Volume 3): 1085 1090 Farnham, UK: British CropProtection Council.	Efforts underway within Monsanto Company to further reduce the unintended impacts to the water environment from the use of its crop protection products are described. Examples ofthese efforts include monitoring and other surveillance activities, investigation and remediation, new farming practices, training and prevention
12.	Gilmore, D W	1997	Ecosystem management a needs driven, resource use philosophy	Forestry Chronicle. 1997; 73(5): 560-564	Ecosystem management is a resource use concept that has evolved as a result of increasing public awareness and society's concerns regarding the future availability of goods and services from the forest resource. This paper reviews the evolution of the ecosystem management philosophy and its use as an approach toward holistic resource stewardship, particularly of forests in North America, to meet the ecological, social and economic needs of society.
13.	Ribaudo, M; Caswell, M F	1999	Environmental regulation in agriculture and the adoption of environmental technology.	Flexible incentives for the adoption of environmental technologies in agriculture. 1999; 7 -25, Dordrecht, Netherlands: Kluwer Academic Publishers.	The US Department of Agriculture has a long history of promoting natural resource stewardship through various forms of voluntary assistance such as education, technical assistance and cost sharing. Water quality programmes that have arisen from the Clean Water Act and state water quality protection laws generally force farmers through command and control or other incentive mechanisms to adopt alternative management practices.
14.	Holsman, R H	2000	Goodwill hunting? Exploring the role of hunters asecosystem stewards.	Wildlife Society Bulletin. 2000; 28(4): 808-816, Bethesda, USA: Wildlife Society.	The expanded role of wildlife management and conservation in addressing diverse social and ecological goals necessitates an examination of the role that hunters have served and can serve in ecosystem stewardship. The notion of ecological stewardship is presented as a moral norm a personal ethic of right and wrong that influences our attitudes and behaviours toward conservation issues and one that requires consideration of the impacts of our lifestyle choices on environmental quality.

Sl. No.	Author	Publishing Year	Title	Source	Abstract
15.	Wander, M M; Drinkwater, L E	2000	Fostering soil stewardship through soil quality assessment	Applied Soil Ecology. 2000; 15(1): 61-73	Educational materials that highlight soil contributions to farm, landscape and global functioning coupled with dialogue between practitioners, scientists, and policy makers can communicate the importance of soil quality to sustainability. Successful soil quality efforts will relate soil properties to soil function in a way that fosters stewardship among individuals and builds public support for policies that promote soil management to ensure agriculture, industry, and the natural environment are sustained. Strategies and priorities are expected to vary according to audience, land use constraints and the intended scale of application.
16.	Lichtenberg, E	2000	Agriculture and the environment.	Working Paper. Department of Agricultural and Resource Economics, University of Maryland. 2000; (00/15): 111 pp. P College Park, USA: University of Maryland.	The source of environmental quality problems and the strength of incentives for resource stewardship that may incidentally induce farmers to protect environmental quality and environmental policy design under the features that are pervasive in agriculture: (1) heterogeneity caused by resource variability; and (2) uncertainty and also the effects of interactions between agricultural, environmental, and resource policies are important implications to design policy distinctive nature of environmental quality problems in agriculture, an industry based on the extraction of highly variable natural resources under stochastic conditions.
17.	Rotterdam, Netherlands: A.A. Balkema.	2000	Reclaimed land:erosion control, soils and ecology.	Reclaimed land: erosion control, soils and ecology.2000; 385	The problems of sustaining the quality of land that has already been reclaimed is possible through minimizing the long term liability of reclaimed land; a strategy for determining acceptable sediment yield for reclaimed mine lands; rainfall erosion on opencast coal mine lands ecological perspective; erosion control principles and some technical options; soil stewardship on reclaimed coal lands;
18.	Murombedzi, J	2001	Committees, rights, costs & benefits: naturalresource stewardship & community benefits in Zimbabwe's CAMPFIRE programme.	African wildlife and livelihoods: the promise and performance of community conservation. 2001; 244-255 Oxford, UK: James Currey Ltd.	Zimbabwe has devolve ownership of wildlife to people in the communal areas, and thereby promoted stewardship of wildlife through the production of benefits for the participating communities. In this regard natural resource management, household accumulation of CAMPFIRE revenues, and local conceptualizations of 'development' in the Masokaarea were taken into account.

Sl. No.	Author	Publishing Year	Title	Source	Abstract
19.	Carruthers, D V	2001	The politics and ecology of indigenous folk art in Mexico	Human Organization. 2001; 60(4): 356-366 ,Oklahoma City, USA: Society for Applied Anthropology	Many indigenous art forms demonstrate characteristics associated with triumphs in agroecology, social forestry, and other rural initiatives, i.e., community revitalization, resource stewardship, traditional ecological knowledge, identity, biological and cultural diversity, and grassroots participation. However, field research in rural Michoacan and Oaxaca presents a sobering counterpoint, finding most peasant artisans entangled, both as agents and victims, in worrisome patterns of degradation, depletion, and exploitation. Global and domestic market failures, predatory bossism, embedded inequities, lack of cooperative success, and political exclusion largely undermine in practice the ecological potential of indigenous art.
20.	Cooke, R J; Park, J R	2001	The use of a decision support tool to assess the financial viability of selected Countryside Stewardship options.	Farm Management. 2001; 11(3): 193-204 Reading, UK: Institute of Agricultural Management, Farm Management Unit, Reading University.	The Countryside Stewardship Scheme offers financial incentives to farmers and landowners to improve the conservation and amenity value of English farmland. These payments often impact on 'core business' income and therefore entry decision requires careful financial appraisal. A decision support tool (the Countryside Stewardship Farm Assessor) was constructed for use by farmers considering entry to the scheme.
21.	Delate, K	2002	Using an agroecological approach to farming systems research.	HortTechnology . 2002; 12(3): 345-354 Alexandria, USA: American Society for Horticultural Science.	Systems theory holds that biological systems, such as agroecosystems, consist of integrated units of people, plants, animals, soil, insects and microorganisms, and each subsystem provides feedback to the other. The goal of each long term agroecological research (LTAR) was to examine the short and long term physical, biological, and socioeconomic effects of organic and conventional farming systems. By establishing long term experiments, longer crop rotations, typical of organic farms, provide yield stability, improve plant protection, and enhance soil health and economic benefits compared to conventional systems with shorter rotations and greater off farm inputs kind of hypothesis was tested.. Feedback from the local farm associations that are responsible for farm stewardship and farm finances is inherent in the LTAR process.

Sl. No.	Author	Publishing Year	Title	Source	Abstract
22.	Letey, J; Sojka, R E; Upchurch, D R; Cassel, D K; Olson, K R; Payne, W A; Petrie, S E; Price, G H; Reginato, R J; Scott, H D; Smethurst, P J; Triplett, G B	2003	Deficiencies in the soil quality concept and its application	Journal of Soil and Water Conservation Ankeny. 2003; 58(4): 180-187, Ankeny, USA: Soil and Water Conservation Society.	The alternative view that emphasis on known principles of soil management is a better expenditure of limited resources for soil stewardship than developing and deploying subjective indices which fail to integrate across the necessary spectrum of management outcomes. Emphasis should be directed toward using available technical information to motivate and educate farmers on management practices that optimize the combined goals of high crop production, low environmental degradation, and a sustained resource.
23.	Miller, F P	2003	Natural resources management and the second copernican revolution.	Journal of Natural Resources and Life Sciences Education. 2003; 32: 43-51, Madison, USA:	The future well being of humanity is inexorably linked to our understanding of the natural world and to its sustainable management. Agriculture and managed renewable natural resources account for >90% of the nation's terrestrial ecosystems. The management and stewardship of
				American Society of Agronomy.	these vast ecosystems along with the ecological goods and services derived therefrom are huge and noble responsibilities. Natural resources management and ecosystem stewardship must be ecology based, including the human dimensions of ecosystem manipulation and natural resource use. Until economic systems recognize the value of environmental investments and marketability of ecological services, adoption of sustainable ecosystem management will be short circuited by rational economic decision making under current economic systems and personal values on the part of those responsible for managing these lands.
24.	S K Acharjee, U.Ghosh, A.Biswas and A.K.Murah	2004	A study on the percseived benefits of farm forestry in terms of socio economic and motivational correlates in Coochbehar district of West Bengal.	"The Indian Foretser", Vol.130, No.3, March 2004.	The role of women in promoting and supporting farm forestry, directly and indirectly, is unquestionably important and essential .Operational holding, land under farm forestry, women's participation and information use index substantively influenced the consequent variable, perceived benefits of farm forestry.

Sl. No.	Author	Publishing Year	Title	Source	Abstract
25.	Landers, J N	2004	Zero tillage challenges and solutions in Brazil.	Proceedings VII World Soybean Research Conference, IV International Soybean Processing and Utilization Conference, III Congresso Brasileiro de Soja Brazilian Soybean Congress, Foz do Iguassu, PR, Brazil, 29 February 5 March, 2004. 2004; 274-281 Londrina, Brazil: Brazilian Agricultural Research Corporation, National Soybean Research Center	Zero Tillage requires a holistic approach towards crop rotations, biological controls and environmental impacts, where farmer led research priorities lead to quicker and more cost effective results. The public image of the farmer as "steward of the land" is crucial to receiving environmental services payments. Farmers and their organizations can contribute considerably to research, especially if they have some effective control in deciding research priorities.
26.	Stonehouse, D P	2004	Sustainability issues in the agri food sector in Ontario, Canada.	Journal of Sustainable Agriculture. 2004; 23(3): 109-124, Binghamton, USA: Food Products Press.	Expanding urbanization/industrialization in southern Ontario is out-competing agriculture for use of some of Canada's best farmland. On the remaining farmland, there has been a trend to consolidate into fewer and larger enterprises, to specialize production lines, to mechanize and automate, to increase usage of imported synthetic inputs, all at the expense of environmental protection, and natural resource stewardship and Sustainability of the agrifood system should include the dimension of rural farm community viability, organic farming,reduced input farming , farmer education and extension assistance, financial assistance, cross compliance measures, and compulsion backed by litigation and penalties.
27.	Franklin, S B; Kupfer, J A;	2004	Biotic diversity of Natchez Trace Brazil.	Environmental Monitoring and Assessment. 2004; International Soybean Processing and Utilization Conference, III Congresso Brasileiro de Soja Brazilian Soybean Congress, Foz do Iguassu, PR, Brazil, 29 February 5 March, 2004. 2004; 274-281 Londrina, Brazil: Brazilian Agricultural Research Corporation, National Soybean Research Center	A test was done to check the functional ability of ecological land types (ELTs) for biodiversity stewardship. Woody and cost effective results. The public image of the farmer as "steward of the land" is crucial to receiving environmental services payments. Farmers and their organizations can contribute considerably to research, especially if they have some effective control in deciding research priorities.

Sl. No.	Author	Publishing Year	Title	Source	Abstract
26.	Stonehouse, D P	2004	Sustainability issues in the agri food sector in Ontario, Canada.	Journal of Sustainable Agriculture. 2004; 23(3): 109-124, Binghamton, USA: Food Products Press.	Expanding urbanization/industrialization in southern Ontario is outcompeting agriculture for use of some of Canada's best farmland. On the remaining farmland, there has been a trend to consolidate into fewer and larger enterprises, to specialize production lines, to mechanize and automate, to increase usage of imported synthetic inputs, all at the expense of environmental protection, and natural resource stewardship and Sustainability of the agrifood system should include the dimension of rural farm community viability, organic farming ,reduced input farming , farmer education and extension assistance, financial assistance, cross compliance measures, and compulsion backed by litigation and penalties.
27.	Franklin, S B; Kupfer, J A; Grubaugh, J W; Kennedy, M L	2004	Biotic diversity of Natchez Trace State Forest, Western Tennessee.	Environmental Monitoring and Assessment. 2004; 93(1/3): 31-54, Dordrecht, Netherlands: Kluwer Academic Publishers.	A test was done to check the functional ability of ecological land types (ELTs) for biodiversity stewardship. Woody and herbaceous flora as well as herpetofauna and avifauna communities had substantial differences between upland and lowland forests. However, none of the faunal communities distinguished among upland (ELTs). In addition, herbaceous taxa also failed to distinguish upland ELTs. Results suggest the present use of ELTs at Natchez Trace State Forest (NTSF) will not be a helpful guide to land stewardship focusing on biodiversity. The disturbance history of the forest and the mobility of animals are given as potential explanations for a poor relationship between abiotic ELTs and the resident biota.
28.	Manoj Kumar Patra, Sankar Kumar Acharjee and Susanta Kumar Chakraborty.	2005	Conservation categories of siluroid fishes in North East Sundarbans, India.	Biodiversity and Conservation (2005) 14 : 1863-1876, @ Springer 2005	The erosion of local fish resources has both immediate and long term impacts. Modernized agriculture and expanding urbanization have not only polluted the surroundings but also threatened the existence of key fisheries taxa.
29.	Leaman, D J	2006	Sustainable wild collection of medicinal	Medicinal and aromatic plants: agricultural, commer	Uncounted additional species are used in the growing cosmetics and botanicals industries. The great majority of these materials is provided

Sl. No.	Author	Publishing Year	Title	Source	Abstract
			and aromatic plants: development of an international standard	cial, ecological, legal, pharmacological and social aspects. 2006; 97-107 Heidelberg, Germany: Springer Verlag GmbH.	by collection from wild resources. Building on successes in the development of sustainable harvest and resource stewardship standards in different sectors such as timber, non wood forest products, organic agriculture and marine aquarium fish, an initiative has recently been launched to develop an international standard and criteria for good practice in the sustainable wild collection of medicinal and aromatic plants.
30.	Mitchell, D P; Norton, T W; Grenfell, R I; Woodgate, P W	2007	Barriers to implementation of natural resource management at the landholding level in the intensive agricultural zones of Southern Australia.	Australasian Journal of Environmental Management. 2007; 14(2): 103-110 ,Melbourne, Australia: Environment Institute of Australia and New Zealand. URL: http://www.eianz.org/	The extent to which past natural resource management (NRM) investment in Australia has actually improved the condition of natural resources is unclear. Many lessons remain to be drawn that can benefit NRM and help empower decision makers responsible for resource stewardship. Lack of information to support decision making, the small returns on agriculture in marginal areas, poor allocation of funding, reluctance to release information and a lack of whole of government approach to management were seen to be important barriers.
31.	Davies, B B; Hodge, I D	2007	Exploring environmental perspectives in lowland agriculture: A Q methodology study in East Anglia, UK.	Ecological Economics. 2007; 61(2/3): 323-333 Oxford, UK: Elsevier. URL: http://www.sciencedirect.co m/science?_ob=ArticleUR L&_udi=B6VDY 4JWFH1G 1	Understanding how farmers conceive of their environmental rights and responsibilities is important in helping to design agri environmental measures that can complement, rather than conflict, with the underlying normative assumptions that farmers themselves hold. Through Q analysis identifies five distinct perspectives on notions of agricultural stewardship were identified termed as Environmentalists, Progressives, Commodity Conservationists, Jeffersonians and Yeomen. It was found that Q methodology can be a valuable way of demonstrating the nature of the mental frameworks of actors in a particular context and this enables us to formulate some important questions regarding the motivations of land managers in the context of a rapidly changing rural policy.
32.	Campbell, A	2008	Food, energy, water: conflicting insecurities (and the rare win wins offered by soil stewardship).	Journal of Soil and Water Conservation Ankeny. 2008; 63(5): 149A-151A Ankeny, USA: Soil and Water Conservation Society. URL: http://www.swcs.org	Soil stewardship offers a crucial key to unlock the apparent tradeoffs between food security, energy security, water security and managing runaway climate change. Improving soil stewardship is fundamental to the world being able to feed itself while improving land, water and energy productivity, and reducing greenhouse gas emissions.

Sl. No.	Author	Publishing Year	Title	Source	Abstract
33.	Overloop, P J van; Miltenburg, I J; Clemmens, A J; Strand, R J	2008	Identification of pool characteristics of irrigation canals.	Proceedings of the World Environmental and Water Resources Congress, Honolulu, Hawai' i, USA, 12-16 May, 2008. 2008; 118, Reston, USA: American Society of Civil Engineers.	In the Western part of the United States of America, more and more irrigation canal networks are being modernized, which often includes some form of real time automatic water level control. The objective of the controller is to maintain the water levels in the canal pools at a given target level. The control loop is repeated at a fixed time interval. To attenuate noise and undesired signal contents, filtering of the water level measurements is required. The reasons that irrigation districts decide to automate their canals include: improved water delivery service, improved efficiency of water delivery, improved utilization of labor, and water resource stewardship.
34.	Lowthe Thomas, S C; Clarke, A; Clapham, S J; Morgan, M; Semere, T; Slater, F M; Smith, R	2008	Reed canary grass a versatile energy crop for Wales?	Aspects of Applied Biology. 2008; (90): 81-86, Wellesbourne, UK: Association of Applied Biologists.	Energy crops are becoming increasingly important for biomass production and farmland diversification and, in Wales, are being considering for integration into agricultural stewardship schemes under an on going review of land management schemes. Trials suggested that reed canary grass crops are economically competitive with other upland land uses, have comparable yields to other energy grasses and have alternative non energy markets. The biodiversity of the crop and potential of the crop for environmental protection are discussed in the context of growing the crop as a low input system.
35.	Grigg, N S	2008	Integrated water resources management: balancing	Water International. 2008; 33(3): 279-292, Carbondale, USA: Interna	Coordination across geographic areas is recognized as a special challenge of integration and emerging efforts toward integration in some countries will bear close scrutiny to
			views and improving practice.	tional Water Resources Association (IWRA). URL: http://pdfserve.informaworl d.com/216710_901698885 .pdf	measure their success. Regardless of the ultimate success in promoting the concept of integrated water resources management, it communicates to stakeholders the notion that water decisions require balance and the promotion of shared values, such as conservation and the alleviation of poverty. If institutional barriers and confusion about the concept are overcome, actual improvements in water stewardship may result.

Sl. No.	Author	Publishing Year	Title	Source	Abstract
36.	Fayram, A H; Schenborn, D A; Hennessy, J M; Nate, N A; Schmalz, P J	2009	Exploring the conflict between broad scale and local inland fisheries management: the risks to agency credibility.	Fisheries Bethesda. 2009; 34(5): 232-236 PB: Bethesda, USA: American Fisheries Society. URL: http://www.fisheries.org/afs /docs/fisheries/fisheries_34 05.pdf	Resource stewardship requires solid fisheries science and long term goals. Agency responsiveness requires a willingness to work with local interests on both long and short term needs within the framework of what is biologically possible. Finally, agency accountability for results includes the obligation to quantitatively monitor, evaluate, and communicate the consequences of management decisions to stakeholders and decision makers.
37.	Ahnstrom, J; Hockert, J; Bergea, H L; Francis, C A; Skelton, P; Hallgren, L	2009	Farmers and nature conservation: what is known about attitudes, context factors and actions affecting conservation?	Renewable Agriculture and Food Systems. 2009; 24(1): 38-47,Cambridge, UK: Cambridge University Press. URL: http://journals.cambridge.or g/action/displayJournal?jid =RAF	Farmers operate in a tight financial situation, and in parts of the world they are highly dependent on government subsidies, and cannot afford to risk losing that support. Use of conservation practices is most effective when these are understood in the context of the individual farm, and decisions are rooted in land and resource stewardship and long term concerns about health of the farm and the soil. As new agri environmental schemes are planned, agricultural development specialists need to recognize the complexity of farmer attitudes, the importance of location and individual farmer circumstances, and the multiple factors that influence decisions.
38.	Lai PoHsin; Sorice, M G; Nepal, S K; Cheng ChiaKuen	2009	Integrating social marketing into sustainable resource management at Padre Island National Seashore: an attitude based segmentation approach.	Environmental Management. 2009; 43(6): 985-998 New York, USA: Springer Verlag. URL: tent/05186h401164j685/?p=e1ad-747f25964a318dc1198904c1cbad&pi=3 http://www.springerlink.co m/con	National Park Service (NPS), is associated with the task with the mission to provide open access for quality outdoor recreation and maintain the ecological integrity of the park system. In addition to management practices of education and restrictions, building a sense of natural resource stewardship among visitors facilitated the NPS ability to react to this task.

Sl. No.	Author	Publishing Year	Title	Source	Abstract
39.	Lertzman, K	2009	The paradigm of management, management systems, and resource stewardship	Journal of Ethnobiology. 2009; 29(2): 339-358 Denton, USA: Society of Ethnobiology. URL: http://www.bioone.org/ doi/ abs/10.2993/0278 0771 29.2.339	Reframing the concept from management to management system provides a tool for better understanding how social and ecological dynamics act as coupled drivers of managed ecosystems. Seen from this perspective, there are strong parallels between the traditional resource management systems of indigenous peoples and western science based management systems. Stewardship is a western concept which resonates with the foundations of traditional resource management systems. Both systems of management can be understood on gradients of human influence on ecosystems and of management intensity. Achieving an integrated understanding of the coupled dynamics of social and ecological systems is a central challenge for both managers and for researchers
40.	Atari, D O A; Yiridoe, E K; Smale, S; Duinker, P N	2009	What motivates farmers to participate in the Nova Scotia environmental farm plan program? Evidence and environmental policy implications.	Journal of Environmental Management. 2009; 90(2): 1269-1279 Amsterdam, Netherlands: Elsevier, URL: http://www.sciencedirect.co m/science/journal/0301479 7	To understand the factors that affect farmer participation in the relatively new Environmental Farm Plan (EFP) program ,Traditional farmer and farm attributes that have contrasting effects in agricultural innovation adoption and conservation management (namely age, and formal education completed), were given least importance, whereas Farm income, years of farming experience, and type of agribusiness managed were of prime importance. Helping to publicize positive farm stewardship practices was reported as the most important reason for participating in the EFP scheme, followed by its use to help improve relationships with non farming neighbours, and to comply with government environmental regulations.
41.	Kay, P; Edwards, A C; Foulger, M	2009	A review of the efficacy of contemporary agricultural stewardship measures for ameliorating water pollution problems of key concern to the UK water industry.	Agricultural Systems. 2009; 99(2/3): 67-75 ,Oxford, UK: Elsevier. URL: http://www.sciencedirect.co m/science/journal/0308521 X	Existing policies require improved water quality and contemporary regulation is encouraging water companies to deal with the problem at source, rather than relying exclusively on 'end of pipe' treatment solutions. Although numerous schemes are available that encourage farmers to adopt environmentally friendly farming practices, uncertainty exists as to the specific impacts of these measures on water quality and, therefore, range of agricultural

Sl. No.	Author	Publishing Year	Title	Source	Abstract
					stewardship measures were listed that have been proven to impact water quality for three pollutant groups of key concern to the UK water industry, namely dissolved organic carbon, nutrients and pesticides.
42.	Siddharta Sankar Mishra, Sankar Kumar Acharjee, Susanta Kumar Chakraborty.	2009	Development of tools for assessing conservation categories of siluroid fishes of fresh water and brackish water wetlands of South West Bengal, India.	Environ Biol Fish (2009) 84 : 395-407, @ Springer.	The lack of information on the occurrence, distribution and status of ichthyofauna of South West Bengal is a big handicap in popularizing and propagating the less known fish varieties in a particular ecosystem for their conservation.
43.	Chapin, F S, III; Carpenter, S R; Kofinas, G P; Folke, C; Abel, N; Clark, W C; Olsson, P; Smith, D M S; Walker, B; Young, O R; Berkes, F; Biggs, R; Grove, J M; Naylor, R L; Pinkerton, E;	2010	Ecosystem stewardship: sustainability strategies for a rapidly changing planet.	Trends in Ecology and Evolution. 2010; 25(4): 241-249, Oxford, UK: Elsevier Ltd.	Ecosystem stewardship is an action oriented framework intended to foster the social ecological sustainability of a rapidly changing planet. Recent developments identify three strategies that make optimal use of current understanding in an environment of inevitable uncertainty and abrupt change: reducing the magnitude of, and exposure and sensitivity to, known stresses; focusing on proactive policies that shape change; and avoiding or escaping unsustainable social ecological traps.
42.	Siddharta Sankar Mishra, Sankar Kumar Acharjee, Susanta Kumar Chakraborty.	2009	Development of tools for assessing conservation categories of siluroid fishes of fresh water and brackish water wetlands of South West Bengal, India.	Environ Biol Fish (2009) 84 : 395-407, @ Springer.	The lack of information on the occurrence, distribution and status of ichthyofauna of South West Bengal is a big handicap in popularizing and propagating the less known fish varieties in a particular ecosystem for their conservation.
43.	Chapin, F S, III; Carpenter, S R; Kofinas, G P; Folke, C; Abel, N; Clark, W C; Olsson, P; Smith, D M S; Walker, B; Young, O R; Berkes, F; Biggs, R; Grove, J M; Naylor, R L; Pinkerton, E; Steffen, W; Swanson, F J	2010	Ecosystem stewardship: sustainability strategies for a rapidly changing planet.	Trends in Ecology and Evolution. 2010; 25(4): 241-249, Oxford, UK: Elsevier Ltd.	Ecosystem stewardship is an action oriented framework intended to foster the social ecological sustainability of a rapidly changing planet. Recent developments identify three strategies that make optimal use of current understanding in an environment of inevitable uncertainty and abrupt change: reducing the magnitude of, and exposure and sensitivity to, known stresses; focusing on proactive policies that shape change; and avoiding or escaping unsustainable social ecological traps.

Sl. No.	Author	Publishing Year	Title	Source	Abstract
44.	Yiridoe, E K; Odwa, D; Atari, A; Gordon, R; Smale, S	2010	Factors influencing participation in the Nova Scotia Environmental Farm Plan Program.	Land Use Policy. 2010; 27(4): 1097-1106, Oxford, UK: Elsevier Ltd. URL: http://www.sciencedirect.co m/science/journal/0264837 7	There is growing interest in Nova Scotia's Environmental Farm Plan (NS EFP) program among farmers and policy makers because of effectiveness of standardized or uniform beneficial management practices in mitigating the negative environmental impacts from agriculture is limited by inherent heterogeneities in agricultural production systems. and there is heightened interest in farmers generating ecological goods and services to society. Regression Analysis suggests that farm characteristics (i.e., farm type, farm size, farm income) and farmer capacity variables (i.e., specialized training and knowledge from EFP (Nova Scotia's Environmental Farm Plan) program information sessions and workshops and on farm steward (ship demonstrations) were significant determinants of environmental farm planning.
45.	Lindenschmidt , K E; Syrenne, G; Harrison, R	2010	Measuring ice thicknesses along the Red River in Canada using RADARSAT 2 satellite imagery.	Journal of Water Resource and Protection. 2010; 2(11): 923-933 Irvine, USA: Scientific Research Publishing. URL: http://www.scirp.org/Journ al/PaperDownload.aspx?Fil eName= =3184 JWARP20101100001_61841300.pdf&paperID	Manitoba Water Stewardship is striving to model the occurrence of ice breakup and simulate the behaviour of ice jamming along the river. An important parameter in ice breakup forecasting is the ice thickness. Good correlations were obtained between the HH backscatter readings and the surveyed ice thicknesses. HV backscatter readings correlate better with fresh snow depth measurements. Additionally, using same sensor incident angle and flight geometry allows ice thickening rate behavior over the course of the winter to be determined.
46.	Ellefson, P V; Kilgore, M A	2010	United States wood based industry: a review of structure and organization.	Staff Paper Series Department of Forest Resources, University of Minnesota. 2010; (206): 238 pp St. Paul, USA: Department of Forest Resources, University of Minnesota. URL:http://www.forestry.umn.ed u/publications/staffpapers/S taffpaper206.pdf	Geographic arrangement, economic and business concentration, diversification, enterprise ownership, market entry, raw material control, research, and environmental and resource stewardship these are the different aspects of wood manufacturing industry which were explored to structure and organization of wood based industry in the USA, which is composed of four major segments or groups, namely: the timber growing segment, timber management support segment, timber harvesting and transport segment, and the wood product manufacturing segment. Among the more significant challenges is the lack of consistent, reliable and comprehensive statistical evidence.

Sl. No.	Author	Publishing Year	Title	Source	Abstract
47.	Deasy, C; Quinton, J N; Silgram, M; Bailey, A P; Jackson, B; Stevens, C J	2010	Contributing understanding of mitigation options for phosphorus and sediment to a review of the efficacy of contemporary agricultural stewardship measures.	Agricultural Systems. 2010; 103(2): 105-109 , Oxford, UK: Elsevier Ltd. URL: http://www.sciencedirect.co m/science/journal/0308521 X	Mitigation Options for Phosphorus and Sediment (MOPS) demonstrated that tramlines can be the major pathway for sediment and nutrient transfer from arable hillslopes, and that although minimum tillage, crop residue incorporation, contour cultivation, and beetle banks also have potential to be cost effective mitigation options, tramline management is the one of the most promising treatments for mitigating diffuse pollution losses, as it was able to reduce sediment and nutrient losses by 72 99% in four out of five site years trialled.
48.	McIntyre, J	2010	Measuring agricultural stewardship: risks and rewards. The case for the stewardship index for specialty crops.	Journal of Agriculture, Food Systems and Community Development. 2010; 1(1): 19-22 , Ithaca, USA: New Leaf Associates. URL: http://www.agdevjournal.co m/attachments/112_JAFSC D_McIntyre_Column_08 10.pdf	Rather than trying to define stewardship and sustainability abstractly, the core goal behind the Stewardship Index is to figure out what specific impacts of food production on people and place matter most to good stewardship and then measure them which will offer information on the Stewardship Index for Specialty Crops, an effort to establish a series of broadly agreed upon "stewardship" metrics for specialty crop supply chains in the USA.
49.	Weber, T C; Allen, W L	2010	Beyond on site mitigation: an integrated, multi scale approach to environmental mitigation and stewardship for transportation projects.	Landscape and Urban Planning. 2010; 96(4): 240-256 , Oxford, UK: Elsevier Ltd. URL: http://www.sciencedirect.co m/science?_ob=ArticleUR L&_udi=B6V91 4YYWRJC 1	Restoration projects of road often are opportunistic, do not consider a broad landscape context, and have a low benefit cost ratio. To identify and evaluate natural resource stewardship opportunities in watersheds that could potentially be affected by construction of a highway bypass, Stewardship activities of which include conservation, restoration, or management of natural resources, would go above and beyond required mitigation. A model of conservation network of high quality wildlife and plant habitat (core areas), large contiguous natural areas (hubs), and linkages to facilitate wildlife movement and gene flow (corridors) was constructed, which ranked elements of this network at multiple scales, and identified high priority areas for conservation and restoration.

Sl. No.	Author	Publishing Year	Title	Source	Abstract
50.	Seidl, R; Fernandes, P M; Fonseca, T F; Gillet, F; Jonsson, A M; Merganicova, K; Netherer, S; Arpaci, A; Bontemps, J D; Bugmann, H; Gonzalez Olabarria, J R; Lasch, P; Meredieu, C; Moreira, F; Schelhaas, M J; Mohren, F	2011	Modelling natural disturbances in forest ecosystems: a review	Ecological Modelling. 2011; 222(4): 903-924,Oxford, UK: Elsevier Ltd.	Considering the increasing importance of disturbances for forest dynamics and ecosystem stewardship under anthropogenic climate change, the latter concepts are crucial tool for understanding and coping with change in forest ecosystems. Current challenges for disturbance modelling in forest ecosystems are thus (i) to overcome remaining limits in process understanding, (ii) to further a mechanistic foundation in disturbance modelling, (iii) to integrate multiple disturbance processes in dynamic ecosystem models for decision support in forest management, and (iv) to bring together scaling capabilities across several levels of organization with a representation of system complexity that captures the emergent behaviour of disturbance regimes.
51.	Thompson, P B	2011	The ethics of soil: stewardship, motivation, and moral framing.	Sustaining soil productivity in response to global climate change: science, policy, and ethics. 2011; 3-1 4, Chichester, UK: John Wiley & Sons. URL: http://onlinelibrary.wiley.co m/doi/10.1002/ 9780470960 257.ch3/summary.	It is possible to conceptualize soil management as a means or instrument for achieving more fundamental ends, such as fulfilling duties to other people or future generations or finding the optimal distribution of goods and evils in the form of human sustenance and environmental insult and to move beyond the instrumentalist understanding of the ethics of soil by considering whether soil stewardship practices contribute to the formation of moral character or community identities in more fundamental way. Such modes of thinking were more commonplace in earlier eras, but have waned in an era when relatively few people engage in active soil management.
52.	Graham, D A; Bonta, M A; Ulloa, R	2011	Cycad conservation, peasant subsistence, and the military coup in Honduras	Society and Natural Resources. 2011; 24(2): 193-200 London, UK: Routledge, Taylor & Francis. URL: http://www.informaworld.c om/ smpp/content~db=all~c ontent=a928331368~-frm=ti tlelink	In rural Honduras, collaborating U.S. and Honduran researchers have elucidated indigenous technical knowledge and customs surrounding local uses for, and conservation of, an endemic cycad DioonmejiaeStandl. & L. O. Williams (Zamiaceae) that is an important food source for the lowest economic stratum of local society. Dissemination of research findings within the region has produced positive outcomes for the cycad's conservation without undermining the usufructuary rights of cycad dependent families. This experience suggests the elements of an effective, low cost approach to promulgating non coercive, community based resource stewardship.

Sl. No.	Author	Publishing Year	Title	Source	Abstract
53.	Williams, J N; Hollander, A D; O'Geen, A T; Thrupp, L A; Hanifin, R; Steenwerth, K; McGourty, G; Jackson, L E	2011	Assessment of carbon in woody plants and soil across a vineyard woodland landscape.	Carbon Balance and Management. 2011; 6(11): (9 November 2011) London, UK: BioMed Central Ltd. URL: http://www.cbm-journal.co m/content/ pdf/1750-0680- 6-11.pdf	Vineyards managed with practices that conserve some fraction of adjoining wildlands yield benefits for increasing overall C stocks and species and habitat diversity in integrated agricultural landscapes. For such complex landscapes, high resolution spatial modeling is challenging and requires accurate characterization of the landscape by vegetation type, physical structure, sufficient sampling, and allometric equations that relate tree species to each landscape. Geographic information systems and remote sensing techniques are useful for integrating the above variables into an analysis platform to estimate C stocks in these working landscapes, thereby helping land managers qualify for greenhouse gas mitigation credits. Correcting these policy shortcomings could create incentives for ecosystem service provision, including C storage, as well as encourage better farm stewardship and habitat conservation.
54.	Fehrenbach, H	2011	How bioenergy related water subsistence, and the military coup in Honduras	Biofuels, Bioproducts and Biorefining. 2011; 5(4): London, UK: Routledge, Taylor & Francis. URL: http://www.informa-world.c om/smpp/ content~db=all~c ontent=a928331368~frm=ti tlelink	Certification schemes like feedstock related schemes such as the Roundtable on Sustainable Palm Oil, the Roundtable on an endemic cycad DioonmejiaeStandl. & L. O. Williams (Zamiaceae) that is an important food source for the lowest economic stratum of local society. Dissemination of research findings within the region has produced positive outcomes for the cycad's conservation without undermining the usufructuary rights of cycad dependent families. This experience suggests the elements of an effective, low cost approach to promulgating non coercive, community based resource stewardship.
53.	Williams, J N; Hollander, A D; O'Geen, A T; Thrupp, L A; Hanifin, R; Steenwerth, K; McGourty, G; Jackson, L E	2011	Assessment of carbon in woody plants and soil across a vineyard woodland landscape.	Carbon Balance and Management. 2011; 6(11): (9 November 2011) London, UK: BioMed Central Ltd. URL: http://www.cbm-journal.co m/content/ pdf/1750-0680- 6-11.pdf	Vineyards managed with practices that conserve some fraction of adjoining wildlands yield benefits for increasing overall C stocks and species and habitat diversity in integrated agricultural landscapes. For such complex landscapes, high resolution spatial modeling is challenging and requires accurate characterization of the landscape by vegetation type, physical structure, sufficient sampling, and allometric equations that relate tree species to each landscape. Geographic information systems and remote sensing techniques are useful for integrating the above variables into

Sl. No.	Author	Publishing Year	Title	Source	Abstract
					an analysis platform to estimate C stocks in these working landscapes, thereby helping land managers qualify for greenhouse gas mitigation credits. Correcting these policy shortcomings could create incentives for ecosystem service provision, including C storage, as well as encourage better farm stewardship and habitat conservation.
54.	Fehrenbach, H	2011	How bioenergy related water impacts are considered by certification schemes.	Biofuels, Bioproducts and Biorefining. 2011; 5(4): 464-473 , Chichester, UK: Wiley Blackwell. URL: http://onlinelibrary.wiley.co m/journal/10.1002/(ISSN)1 932 1031	Certification schemes like feedstock related schemes such as the Roundtable on Sustainable Palm Oil, the Roundtable on responsible Soy, the Better Sugarcane Initiative; bioenergy specific schemes such as The Roundtable on Sustainable Biofuel, the international Sustainability and Carbon Certification, the Green Gold Label; forestry related schemes such as the Sustainable Forestry initiative and the Forest Stewardship Council; agriculture related schemes such as the Sustainable Agriculture Network Rainforest Alliance; and water related schemes such as the Alliance for Water Stewardship includes checklists that include criteria and indicators regarding water stress and/or pollution in order to identify sustainable against non sustainable bioenergy production to address the water issue like excessive water consumption, water scarcity and protection of water quality.
55.	Lu, F; Wirth, C	2011	Conservation perceptions, common property, and cultural polarization among the Waorani of Ecuador's Amazon	Human Organization. 2011; 70(3): 233-243 , Oklahoma City, USA: Society for Applied Anthropology. URL: http://www.sfaa.net/ho/	The perception of resource scarcity is more prevalent than before, the main obstacles to fostering resource stewardship among the Waorani are increased tensions between individual rights and collective duties and changing ideas of territorial boundaries. There is a need to give importance towards understanding the role of cultural polarization as a factor in common property management and community conservation in an Amazonian context.
56.	Hoag, D L K; Boone, R B; Keske, C M H	2011	The cost for agriculture to coexist with wildlife in Colorado.	Human Dimensions of Wildlife. 2011; 16(5): 318- 329 Philadelphia, USA: Taylor & Francis, Inc. URL: http://www.informaworld.c om/smpp/title~content=t71 3666717~db=all	Three secondary sources were used to estimate costs incurred by agricultural producers to coexist with wildlife in Colorado, as there is little data, and demonstrates that secondary data sources can provide reliable estimates, and those are: government expenditures, spatial data (e.g., geographic information systems),

Sl. No.	Author	Publishing Year	Title	Source	Abstract
					and published studies. The results inform policy makers and producers alike about the value that agricultural stewardship provides wildlife. Results also provide evidence that secondary data sources are a cost effective, powerful approach that can supplement surveys.
57.	Hancock, D W	2012	Grazing schools improve the sustainability of pasture based livestock enterprises.	Journal of Extension. 2012; 50(5): 5IA-W5 , West Lafayette, USA: Extension Journal Inc. URL: http://www.joe.org/joe/201 2october/iw5.php	Implementing intensive grazing management can improve the three primary aspects of sustainability of enhancing profitability, natural resource stewardship, and quality of life. Participants in Georgia Grazing Schools (GGS) held between 2007 2010 were surveyed and reported increased profitability, with nearly 18.9% and 18.9% reporting an increase in net returns of 20 30% and over 30%, respectively. Respondents also highly rated principles learned and experiences gained at the GGS as aiding them in meeting goals of improved natural resource stewardship and quality of life on their farms
58.	Fischer, J; Dyball, R; Fazey, I; Gross, C; Dovers, S; Ehrlich, P R; Brulle, R J; Christensen, C; Borden, R J	2012	Human behavior and sustainability.	Frontiers in Ecology and the Environment. 2012; 10(3): 153-160 PB: Washington, USA: Ecological Society of America. URL: http://www.esa-journals.org/ doi/full/10.1890/110079	Unsustainable behaviors result from a vicious cycle, where traditional market and state institutions reinforce disincentives for more sustainable behaviors while, at the same time, the institutions of civil society lack momentum to effectively promote fundamental reforms of those institutions. Achieving more sustainable behaviors requires this cycle to be broken. It is to contribute to social change through involvement in initiatives like the Ecological Society of America's Earth Stewardship Initiative or the nascent Millennium Alliance for Humanity & the Biosphere.
59.	Rome, Italy: Food and Agriculture Organization of the United Nations (FAO).	2012	FAO Workshop on International Guidelines for Securing Sustainable Small Scale Fisheries, Rome, Italy, 7 10 February 2012.	FAO Fisheries and Aquaculture Report. 2012; (1004): i vii, 1-44	The small scale fisheries have importance as a contributor to poverty alleviation, food and nutrition security, and economic development. Additional principles, approaches and frameworks to be referred to include equity, economic viability, ecosystem based, holistic and integrated approaches. There is a need to build bridges between different stakeholder visions within the fisheries sector as well as outside to ensure coherence. A new vision on how to ensure access for small scale fisheries communities and

Sl. No.	Author	Publishing Year	Title	Source	Abstract
					giving them resource stewardship responsibilities may be needed, reflecting the need for combining livelihood security and environmental sustainability as one cannot be achieved without the other.
60.	Allison, E H; Ratner, B D; Asgard, B; Willmann, R; Pomeroy, R; Kurien, J	2012	Rights based fisheries governance: from fishing rights to human rights.	Fish and Fisheries. 2012; 13(1): 14-29 , Oxford, UK: Wiley Blackwell. URL: http://onlinelibrary. wiley.co m/journal/10.1111/ (ISSN)1 467 2979	In small scale developing country fisheries in particular, basing the case for fishery governance reform on assumed economic incentives for resource stewardship is insufficient when there are other sources of insecurity in people's lives that are unrelated to the state of fishery resources. The more secure, less vulnerable fishers make more effective and motivated fishery managers in the context of participatory or rights based fisheries governance, and insecurity among fishers living in poverty can be most effectively addressed by social and political development that invokes the existing legal framework supporting the Universal Declaration of Human Rights. Embedding the fisheries governance challenge within a broader perspective of human rights enhances the chances of achieving both human development and resource sustainability outcomes in small scale fisheries of developing countries.
59.	Rome, Italy: Food and Agriculture Organization of the United Nations (FAO).	2012	FAO Workshop on International Guidelines for Securing Sustainable Small Scale	FAO Fisheries and Aquaculture Report. 2012; (1004): i vii, 1-44	The small scale fisheries have importance as a contributor to poverty alleviation, food and nutrition security, and economic development. Additional principles, approaches and frameworks to be referred to include equity, economic viability, ecosys tem based, holistic and integrated
			Fisheries, Rome, Italy, 7 10 February 2012.		approaches. There is a need to build bridges between different stakeholder visions within the fisheries sector as well as outside to ensure coherence. A new vision on how to ensure access for small scale fisheries communities and giving them resource stewardship responsibilities may be needed, reflecting the need for combining livelihood security and environmental sustainability as one cannot be achieved without the other.

Sl. No.	Author	Publishing Year	Title	Source	Abstract
60.	Allison, E H; Ratner, B D; Asgard, B; Willmann, R; Pomeroy, R; Kurien, J	2012	Rights based fisheries governance: from fishing rights to human rights.	Fish and Fisheries. 2012; 13(1): 14-29 , Oxford, UK: Wiley Blackwell. URL: http://onlinelibrary.wiley.co m/journal/10.1111/ (ISSN)1 467 2979	In small scale developing country fisheries in particular, basing the case for fishery governance reform on assumed economic incentives for resource stewardship is insufficient when there are other sources of insecurity in people's lives that are unrelated to the state of fishery resources. The more secure, less vulnerable fishers make more effective and motivated fishery managers in the context of participatory or rights based fisheries governance, and insecurity among fishers living in poverty can be most effectively addressed by social and political development that invokes the existing legal framework supporting the Universal Declaration of Human Rights. Embedding the fisheries governance challenge within a broader perspective of human rights enhances the chances of achieving both human development and resource sustainability outcomes in small scale fisheries of developing countries.
61.	Sanjines, P B	2012	Opportunity analysis of payment for ecosystem services: policy design and implementation for coffee agroforestry systems in Costa Rica.	Tropical Resources: Bulletin of the Yale Tropical Resources Institute. 2012; 31: 56-66 ,New Haven, USA: Yale School of Forestry and Environmental Studies.	The final recommendations from the study shed light on new avenues to design schemes that will foster environmental stewardship, ecosystem service provision, and collaboration among stakeholders. Only through collaborative management and the development of strong support networks will motivate the farmers and other small scale agricultural landowners to be effectively engaged in these mechanisms.
62.	Hartmann, M; Howes, C G; VanInsberghe, D; Yu, H; Bachar, D; Christen, R; Nilsson, R H; Hallam, S J; Mohn, W W	2012	Significant and persistent impact of timber harvesting on soil microbial communities in northern coniferous forests.	ISME Journal. 2012; 6(12): 2199 2218, London, UK: Nature Publishing Group. URL: http://www.nature.com/ism ej/journal/v6/n12/full/ismej 201284a.html	Forest ecosystems have integral roles in climate stability, biodiversity and economic development. Soil stewardship is essential for sustainable forest management. Organic matter (OM) removal and soil compaction are key disturbances associated with forest harvesting, but their impacts on forest ecosystems are not well understood. The results establish a comprehensive inventory of bacterial and fungal community composition in northern coniferous forests and demonstrate the long term response of their structure to key disturbances associated with forest harvesting.

Sl. No.	Author	Publishing Year	Title	Source	Abstract
63.	Kay, P; Grayson, R; Phillips, M; Stanley, K; Dodsworth,	2012	The effectiveness of agricultural stewardship for improving water quality at the	Journal of Hydrology Amsterdam. 2012; 422/423: 10 16 , Oxford, UK: Elsevier Ltd. URL:	Agriculture is estimated to be responsible for 70% of nitrate and 30 50% of phosphorus pollution, contributing to ecological and water treatment problems. Despite the fact that significant gaps remain in our understanding, it is known that agricultural stewardship can be highly effective in controlling
	A; Hanson, A; Walker, A; Foulger, M; McDonnell, I; Taylor, S		catchment scale: experiences from an NVZ and ECSFDI watershed.	http://www.sciencedirect.co m/science/article/pii/ S0022 16941100878X	water pollution at the plot and field scales. Agricultural stewardship measures have the potential to improve water quality at the catchment scale but that voluntary schemes with insufficient financial reward or regulatory pressure are unlikely to be successful.
64.	Krasny, M E; Tidball, K G	2012	Civic ecology: a pathway for Earth Stewardship in cities.	Frontiers in Ecology and the Environment. 2012; 10(5): 267- 273 ,Washington, USA: Ecological Society of America. URL: http://www.esa-journals.org/ doi/ full/10.1890/110230	In an increasingly urban society, city residents are finding innovative ways of stewarding nature that integrate environmental, community, and individual outcomes. These urban civic ecology practices including community gardening, shellfish reintroductions, tree planting and care, and "friends of parks" initiatives to remove invasive and restore native species generally begin as small, self organized efforts after a prolonged period of economic and environmental decline or more sudden major disruptions, such as earthquakes, hurricanes, and conflict.
65.	Rozzi, R; Armesto, J J; Gutierrez, J R; Massardo, F; Likens, G E; Anderson, C B; Poole, A; Moses, K P; Hargrove, E; Mansilla, A O; Kennedy, J H; A; Hanson, A; Walker, A; Foulger, M; McDonnell, I; Taylor, S	2012	Integrating ecology and environmental ethics: Earth stewardship in the Southern end of the Americas. catchment scale: experiences from an NVZ and ECSFDI watershed.	SO: BioScience . 2012; 62(3): 226- 236 ,Washington, USA: American Institute of Biological Sciences. URL: http://www.bioone.org/doi/ full/10.1525/ bio.2012.62.3. 4 http:// www.sciencedirect.co m/ science/article/pii/S0022 16941100878X	A field environmental philosophy methodology was developed by the Chilean Long Term Ecological Research LTSER network to integrate ecological sciences and environmental ethics into graduate education and biocultural conservation. This approach broadens the prevailing economic spectrum of social dimensions considered by LTSER programs and helps foster bioculturally diverse forms of Earth stewardship water pollution at the plot and field scales. Agricultural stewardship measures have the potential to improve water quality at the catchment scale but that voluntary schemes with insufficient financial reward or regulatory pressure are unlikely to be successful.

Sl. No.	Author	Publishing Year	Title	Source	Abstract
64.	Krasny, M E; Tidball, K G	2012	Civic ecology: a pathway for Earth Stewardship in cities.	Frontiers in Ecology and the Environment. 2012; 10(5): 267- 273 ,Washington, USA: Ecological Society of America. URL: http://www.esa-journals.org/ doi/ full/10.1890/110230	In an increasingly urban society, city residents are finding innovative ways of stewarding nature that integrate environmental, community, and individual outcomes. These urban civic ecology practices including community gardening, shellfish reintroductions, tree planting and care, and "friends of parks" initiatives to remove invasive and restore native species generally begin as small, self organized efforts after a prolonged period of economic and environmental decline or more sudden major disruptions, such as earthquakes, hurricanes, and conflict.
65.	Rozzi, R; Armesto, J J; Gutierrez, J R; Massardo, F; Likens, G E; Anderson, C B; Poole, A; Moses, K P; Hargrove, E; Mansilla, A O; Kennedy, J H;Willson, M; Jax, K; Jones, C G; Callicott, J B; Arroyo, M T K	2012	Integrating ecology and environmental ethics: Earth stewardship in the Southern end of the Americas.	SO: BioScience . 2012; 62(3): 226- 236 ,Washington, USA: American Institute of Biological Sciences. URL: http://www.bioone. org/doi/ full/10.1525/ bio.2012.62.3. 4	A field environmental philosophy methodology was developed by the Chilean Long Term Ecological Research LTSER network to integrate ecological sciences and environmental ethics into graduate education and biocultural conservation. This approach broadens the prevailing economic spectrum of social dimensions considered by LTSER programs and helps foster bioculturally diverse forms of Earth stewardship.
66.	Holmes, M C C; Jampijinpa, W	2013	Law for country: the structure of Warlpiri ecological knowledge and its application to natural resource management and ecosystem stewardship.	SO: Ecology and Society. 2013; 18(3): Art. 19,Waterloo, Canada: Resilience Alliance.	Indigenous Ecological Knowledge (IEK) is deeply encoded in social processes. Our research shows that from an Indigenous perspective, IEK is a way of living whose core aim is to sustain the healthy functioning of people and country through relationships of reciprocity
67.	Mason, N	2013		ODI Working Paper. 2013; (363): viii + 48 pp London, UK: Overseas Development Institute. URL: http://www.odi.org.uk/ sites/ odi.org.uk/files/ odi assets/publications opinion files/8190.pdf	Multinational corporations (MNCs) are recognizing water as a central business concern, in this regard a new rubric of 'water stewardship' and management of 'shared water risks' has emerged. This encompasses discrete attempts to reduce water dependency and wastewater impacts in plant operations. A simple schematic is adopted, whereby potential contributions by MNCs to three necessary but all too often lacking conditions for water security namely accountable institutions, transparent information,

Sl. No.	Author	Publishing Year	Title	Source	Abstract
					and equitable and sustainable investments are considered at each of four scales, ranging from MNCs' internal operations, to interactions with communities in the immediate catchment, to the supply chain, and finally engagement at river basin or national level. In each case, an 'area for action' is proposed the objective being to increase the prospects of MNC engagement on water yielding positive results for all, rather than entrenching the influence of a relatively narrow set of interests.
68.	Kang, M S; Banga, S S	2013	Global agriculture and climate change.	Journal of Crop Improvement. 2013; 27(6): 667- 692 ,Philadelphia, USA: Taylor &	A sustainable policy shift toward enhanced food security, preservation of freshwater resources, prevention of soil degradation, and maintenance of biological diversity and ecosystems remain the hallmark of mitigation
				Francis,**URL:** http://www.tandfonline.co m/loi/wcim20	strategies. Improved adaptive capacity relative to climate resilient agriculture needs to be integrated with global developmental paths aimed at reducing social inequalities and alleviating poverty. World essentially requires a climate closest to which all forms of life have adapted during their evolution.
69.	Plieninger, T; Bieling, C	2013	Resilience based perspectives to guiding high nature value farmland through socioeconomic change.	Ecology and Society. 2013; 18(4): art20, Waterloo, Canada: Resilience Alliance. URL: http://www.ecologyandsoci ety.org/vol18/iss4/art20/	Using landscape level cases across Europe, the following social ecological system properties and components and their integration into High Nature Value (HNV) farmland management were considered: (1) coupling of social and ecological systems, (2) key variables, (3) adaptive cycles, (4) regime shifts, (5) cascading effects, (6) ecosystem stewardship and collaboration, (7) social capital, and (8) traditional ecological knowledge. The previous conservation efforts for HNV farmland have focused too much on static, isolated, and monosectoral conservation strategies, and that stimulation of resilience and adaptation is essential for guiding HNV farmland through rapid change.
70.	Sayre, N F; McAllister, R R J; Bestelmeyer, B T; Moritz, M; Turner, M D	2013	Earth Stewardship of rangelands: coping with ecological, economic, and political marginality.	Frontiers in Ecology and the Environment. 2013; 11(7): 348- 354 ,Washington, USA: Ecological Society of America.	Earth Stewardship of rangelands must address both livestock management and the broader socio-ecological dynamics that promote land use changes, fragmentation, and degradation. It was specific gradients on which human rangeland systems can be arrayed,

Sl. No.	Author	Publishing Year	Title	Source	Abstract
				URL: http://www.esa-journals.org/ doi/ full/10.1890/120333	including issues of variability, adaptation to disturbance, commercialization, land use change, land tenure security, and effective governance, and we illustrate the gradients' interactions and effects in sites worldwide. The result is a synthetic framework to help in understanding how rangeland Earth Stewardship can be achieved in the face of marginality, globalization, and climate change.
71.	Turner, N J; Plotkin, M; Kuhnlein, H V	2013	Global environmental challenges to the integrity of Indigenous Peoples' food	Indigenous peoples' food systems and well being: interventions and policies	The integrity of Indigenous Peoples' food systems is intimately connected to the overall health of the environment. Recent declines in many aspects of environmental quality, from loss of biodiversity to environmental
			systems	for healthy communities. 2013; 23 38	contamination, have combined with social, economic, political and cultural factors to threaten the health and well being of Indigenous Peoples, and ultimately of people everywhere. Reconnecting Indigenous Peoples with their traditional territories, and reversing some of the restrictive regulations against Indigenous Peoples' historical hunting and plant harvesting practices may help to restore and maintain traditional resources. More cooperative arrangements for co management of habitats and resources should be instated. Collaborative research is recommended, such as environmental and other relationships among Indigenous Peoples' cultures, lands and resource stewardship are complemented with supporting work by academic partners.
70.	Sayre, N F; McAllister, R R J; Bestelmeyer, B T; Moritz, M; Turner, M D	2013	Earth Stewardship of rangelands: coping with ecological, economic, and political marginality.	Frontiers in Ecology and the Environment. 2013; 11(7): 348- 354 ,Washington, USA: Ecological Society of America. URL: http://www.esa-journals.org/ doi/ full/10.1890/120333	Earth Stewardship of rangelands must address both livestock management and the broader socio-ecological dynamics that promote land use changes, fragmentation, and degradation. It was specific gradients on which human rangeland systems can be arrayed, including issues of variability, adaptation to disturbance, commercialization, land use change, land tenure security, and effective governance, and we illustrate the gradients' interactions and effects in sites worldwide. The result is a synthetic framework to help in understanding how rangeland Earth Stewardship can be achieved in the face of marginality, globalization, and climate change.

Sl. No.	Author	Publishing Year	Title	Source	Abstract
71.	Turner, N J; Plotkin, M; Kuhnlein, H V	2013	Global environmental challenges to the integrity of Indigenous Peoples' food systems	Indigenous peoples' food systems and well being: interventions and policies for healthy communities. 2013; 23 38	The integrity of Indigenous Peoples' food systems is intimately connected to the overall health of the environment. Recent declines in many aspects of environmental quality, from loss of biodiversity to environmental contamination, have combined with social, economic, political and cultural factors to threaten the health and well being of Indigenous Peoples, and ultimately of people everywhere. Reconnecting Indigenous Peoples with their traditional territories, and reversing some of the restrictive regulations against Indigenous Peoples' historical hunting and plant harvesting practices may help to restore and maintain traditional resources. More cooperative arrangements for co management of habitats and resources should be instated. Collaborative research is recommended, such as environmental and other relationships among Indigenous Peoples' cultures, lands and resource stewardship are complemented with supporting work by academic partners. Ultimately, this will help to maintain and strengthen the resilience of ecosystems and cultural systems, including diverse and healthy food systems.
72.	Rozza, J P; Richter, B D; Larson, W M; Redder, T; Vigerstol, K; Bowen, P	2013	Corporate water stewardship: achieving a sustainable balance.	Journal of Management and Sustainability. 2013; 3(4): 41- 52 ,Toronto, Canada: Canadian Center of Science and Education. URL: http://www.ccsenet.org/jour nal/index.php/jms/article/vi ew/28896/18665	Quantification of community water partnerships (CWP) project that provide water access and sanitation, watershed restoration and protection, and water for productive use, benefits is achieved either through metering or standard methods known and accepted by the engineering, conservation, and social science professions. A 100% sustainable balance is achieved when an enterprise implements a portfolio of locally relevant CWPs that collectively produce an annual volumetric benefit equivalent to the annual volume of consumptive water use for that particular enterprise and can also support efforts by industries that seek to operate in a manner consistent with the United Nations resolution regarding the human right to water and sanitation, as well as those that may seek certification under emerging global water stewardship standards.

Sl. No.	Author	Publishing Year	Title	Source	Abstract
73.	Adhikary, M M; Pradhan, K; Mishra, S	2013	Analysing and ensuring persuasive extension with respect to preventive innovation for sustainable agriculture.	Journal of Crop and Weed. 2013; 9(1): 56-68 ,West Bengal, India: Crop and Weed Science Society.	In the changed agricultural scenario, as concerns about environmental protection, natural resource stewardship, and the India's ability to feed ever growing populations continue to mount, the sustainability of agriculture and natural resources is emerging as a central theme among the public and policymakers alike. Increased environmental awareness and health consciousness promoted the scientists and planners to think and promote an alternate pathway of extension science for sustainable agriculture with protection to environment and human health, called persuasive extension.
74.	Lange, B; Shepheard, M	2014	Changing conceptions of rights to water? An eco socio legal perspective.	Journal of Environmental Law. 2014; 26(2): 215-242,Oxford, UK: Oxford University Press. URL: http://jel.oxfordjournals.org /	On the basis of data an eco socio legal perspective that foregrounds three interpretive frames for understanding how conceptions of rights to water are generated. These are the institutional legal framework of abstraction licensing in England and Wales, perceptions of the natural space which is governed by this legal framework, and, the economic context in which rights to water are exercised
75.	Jat, R A; Sahrawat, K L; Kassam, A H; Friedrich, T	2014	Conservation agriculture for sustainable and resilient agriculture: global status, prospects and challenges.	Conservation agriculture: global prospects and challenges. 2014; 1-25 Wallingford, UK: CABI. URL: http://www.cabi.org/CABeBooks/default.aspx?site=10 7&page=45&LoadModule= PDFHier&BookID=840	Conservation agriculture is the way forward for sustainable agricultural production which include soil and water conservation; soil quality; rainwater use efficiency; nutrient use efficiency; input use efficiency; insect pest, disease and weed dynamics; crop productivity; climate change mitigation and adaptation; benefits at ecosystem level; and farm profitability.
76.	Rockstrom, J; Falkenmark, M; Allan, T; Folke, C; Gordon, L; Jagerskog, A; Kummu, M; Lannerstad, M; Meybeck, M; Molden, D; Postel, S; Savenije, H H G; Svedin, U; Turton, A; Varis, O	2014	The unfolding water drama in the Anthropocene: towards a resilience based perspective on water for global sustainability.	Ecohydrology . 2014; 7(5): 1249-1261 Chichester, UK: Wiley Blackwell. URL: http://onlinelibrary.wiley.co m/journal/10.1002/(ISSN)1 936 0592	The rising water turbulence in the Anthropocene changes the water research and policy agenda, from a water resource efficiency to a water resilience focus. This includes integrated land and water stewardship to sustain wetness dependent ecological functions at the landscape scale and a stronger emphasis on green water management for ecosystem services. A new paradigm of water governance emerges, encouraging land use practices that explicitly take account of the multifunctional roles of water, with adequate attention to planetary freshwater boundaries and cross scale interactions.

Sl. No.	Author	Publishing Year	Title	Source	Abstract
77.	Bardsley, D K; Bardsley, A M	2014	Organising for socio ecological resilience: the roles of the mountain farmer cooperative Genossenschaft Gran Alpin in Graubunden, Switzerland.	Ecological Economics. 2014; 98: 11-21,Oxford, UK: Elsevier Ltd. URL: http://www.sciencedirect.co m/science/article/pii/ S0921 800913003595	Gran Alpin provides a secure premium price for cereal producers in Graubunden linked to the uniqueness of local organic mountain cropping systems, and all the values of local identity, landscape stewardship, biodiversity conservation and regional development that such systems represent .Resilience within the socio ecological system is enhanced as the cooperative exploits evolving forms of collaboration, market niches, and private and public governance relationships to respond implicitly and explicitly to agro ecological, economic and political risks.
78.	Lahiri Dutt, K	2014	Experiencing and coping with change: women headed farming households in the Eastern Gangetic Plains.	ACIAR Technical Reports Series. 2014; (83): 66 pp ,Canberra, Australia: Australian Centre for International Agricultural Research (ACIAR). **URL:** http://www.aciar.gov.au/ tax onomy/term/15	Recommendations for developing projects targeted at women heads of households in the EGP include: the survey suggested incorporation of 'gender mainstreaming' into the design of projects; investigating solutions to labour shortages; recognizing diversity between local contexts and individual circumstances; helping to form community support for women heads of households, to assist them to secure both financial and technical support; involving men in gender equity processes; building capacity of women heads of households; and improving women's access to information, markets and transport.
79.	Stokka, G L; Maddock, R; Dahlen, C; Petry, T; Stoltenow, C	2014	Livestock stewardship: pregnancy checking, why should we recommend it?.	Proceedings of the 46th Annual Conference of the American Association of Bovine Practitioners, Milwaukee, Wisconsin, USA, 19 21 September 2013. 2014; 34-35,Stillwater, USA: American Association of Bovine Practitioners.	The definition of stewardship from Webster's dictionary is "the careful and responsible management of something entrusted to one's care". As livestock stewards, we have a responsibility to manage and utilize the resource of cull animals. Cattle culled from beef and dairy herds fill an important role in meeting the nutritional needs of consumers. With proper care, feeding, and good marketing, there is opportunity to increase the market value of cull cows through weight gain, and a change in the grading category to meet the high standards of our food production system, and to provide a more acceptable product to our consumers.

Sl. No.	Author	Publishing Year	Title	Source	Abstract
80.	A.M.Puste, S.K.Gunri, K. Jana, B. Ray Pramanik, S.K. Acharya, M. Dasgupta and T.K.Maity.	2017	Integrating management for land water ecosystem for augmenting productivity, Income, and sustainable livelihood.	@ Springer Nature Singapore Pte Ltd. 2017, U.K.De et al. (eds.) , Inequality, poverty and development in India, India Studies in Business and Economics.	Among the natural resources, wetland has a key and pivotal role and most integral should be utilized wisely for mankind for its sustainability and their livelihood development. A greater wetland area, especially in north eastern part of the country, is lying vacant, almost without cultivation round the year. This may effectively be utilized through the cultivation of valuable underutilized wetland crops, extinct fish genotypes, etc. with systematically following integrated farming system (IFS).
81.	Manjishtha Bhattacharyya , Sankar kr Acharya and Susanta K. Chakraborty.	2017	Pollinators Unknown: People's perception of native bees in an agrarian district of West Bengal, India, and its implication in conservation.. India, and its implication in conservation..	Tropical Conservation Science , Vol 10: 1-14, sage publication.	Native bees are important pollinators of cultivated and wild plants. But people residing in villages and directly or indirectly associated with agriculture are unaware of the existence of alternate bee pollinators even though they see them every day. Public awareness is essential for implementing pollinator conservation and management campaign and very little effort is needed to educate people on the identity and nature of these alternate bee pollinators occurring in their vicinity. alternate bee pollinators occurring in their vicinity.

3
CHAPTER

THEORETICAL ORIENTATION

Conceptualizing the dynamics of farm stewardship based on conservation This chapter deals with the broadly described conceptual framework used in this study in order to provide a theoretical base for the empirical investigation and guidance for selection of relevant predictor variables.

CONCEPTUAL FRAMEWORK

A conceptual framework is generally constructed to provide the researcher the elements of attainable and researchable problems and ideas supported by methodology, design, prospective analysis, and empirical conclusion. The topic of the present research being Paradigm Shift of Small and Marginal Farmers From Conventional Conservation Practices To Stewardship: The Perception And Possibilities. It is expected that theterminologies and concepts used to frame up the topic need to undergo an analytical text and literature.

There has been a classical confusion between conservation and stewardship whilethe former is centering around maintenance to protect the count of resource and diversity,the latter keeps focusing on mentoring with a purpose to socio ecological resilience. The present conceptual framework has to accommodate and assimilate the texts of differences and similarities between conservation and stewardship. And all these have to be configured within a conceptual framework the research is designing for.

Understanding Change: Conservation and Stewardship

Agricultural production offers both biological and ecological functions to be organized within an ecological setup which again has dovetailed to other ecological Holon's do define, what we may call ecological networking. The networking function helps both ecological and sub ecological functions go together to usher a multi ecologicalfunctionality. The world is undergoing unprecedented changes in many of the factors that determine both its fundamental properties and their influence on society. Throughout human history, people have interacted with and shaped ecosystems for social and economic development (Turner et al. 1990, Redman 1999, Jackson 2001, Diamond 2005). During the last 50 years, however, human activities have changed ecosystems more rapidly and extensively than at any comparable period of human history (Steffen et al. 2004, Foley et al. 2005, MEA 2005d). Earth's climate, for example, is now warmer than at any time in the last 500 (and probably the last 1,300) years (IPCC 2007), in part because of atmospheric accumulation of carbon dioxide (CO_2) released by the burning of fossil fuels. Agricultural development largely accounts for the accumulation of other trace gases that contribute to climate warming.

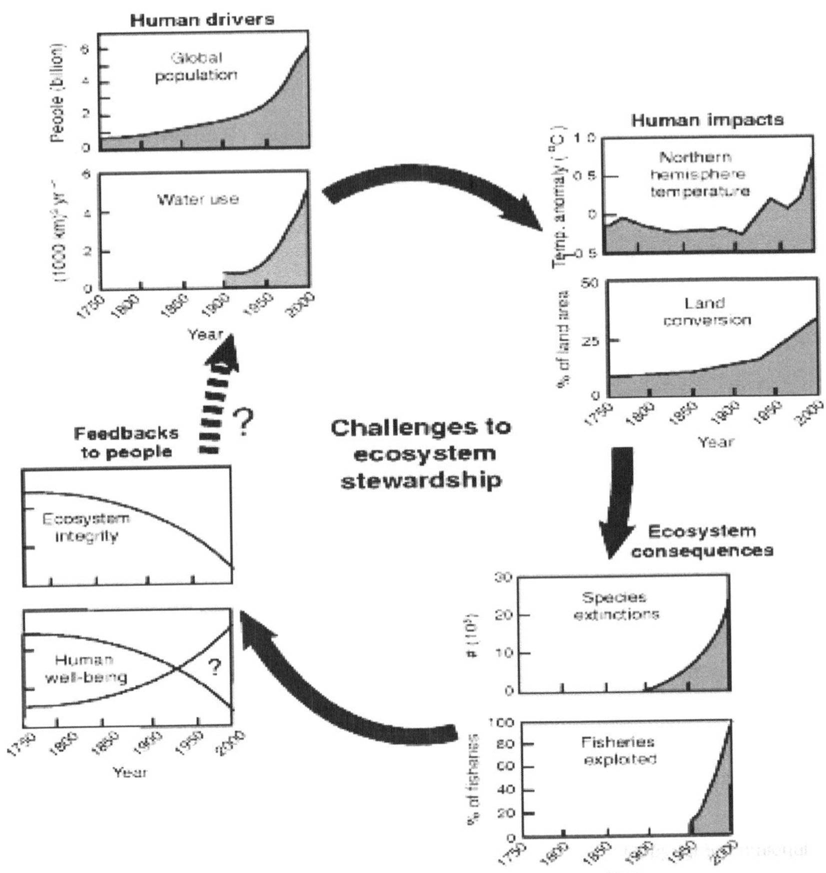

Figure 3.1. Challenges to ecosystem stewardship. Changes in human population and resource consumption alter climate and land cover, which have important ecosystem consequences such as species extinctions and overexploitation of fisheries. These changes reduce ecosystem integrity and have regionally variable effects on human well being, which feeds back to further changes in human drivers. Panel inserts redrawn from Steffen et al. (2004).

Spurt of population across the globe has blatantly has intruded and encroached upon the natural habitats and resources, earlier called the bounties of nature. As human population increases, in part due to improved disease prevention, the increased demand for food and natural resources has led to an expansion of agriculture, forestry, and other human activities, causing large scale land cover change and loss of habitats and biological diversity. About half of the world's population now lives in cities and depends on connections with rural areas worldwide for food, water, and waste processing. In addition, increased human mobility is spreading plants, animals, diseases, industrial products, and cultural perspectives more rapidly than ever before. This increase in global mobility, coupled with increased connectivity through global markets and new forms of communication, links the world's economies and cultures, so decisions in one place often have international consequences.

This globalization of economy, culture, and ecology is important because it modifies the life support system of the planet (Odum 1989), i.e., the capacity of the planet to meet the needs of all organisms, including people. The dramatic increase in the extinction rate of species (100 to 1,000 fold in the last two centuries) indicates that global changes have been catastrophic for many species, although some species, especially invasive species and some disease organisms, have benefited and expanded their ranges. Human society has both benefited and suffered from global changes, with increased food production, increased income and living standards (in parts of the world), improved treatment of many diseases, and longer life expectancy being offset by deterioration in ecosystem services, the benefits that society receives from ecosystems.

More than half of the ecosystem services on which society depends for survival and a good life have been degraded—not deliberately, but inadvertently as people seek to meet their material desires and needs (MEA 2005). Change creates both challenges and opportunities. People have amply demonstrated their capacity to alter the life support system of the planet. With appropriate stewardship, this human capacity can be mobilized to not only repair but also enhance the capacity of Earth's life support system to support societal development. The unique feature of the changes is that they are directional. In other words, they show a persistent trend over time. Many of these trends have become more pronounced since the mid twentieth century and will probably continue or accelerate in the coming decades, even if society takes concerted actions to reduce some rates of change. This situation creates a dilemma in planning for the future because we cannot assume that the future world will behave as we have known it in the past or that our past experience provides an adequate basis to plan for the future. This issue is especially acute for sustainable management of natural resources.

It is no longer possible to manage systems so they will remain the same as in the recent past, which has traditionally been the reference point for resource managers and conservationists. We must adopt a more flexible approach to managing resources— management to sustain the functional properties of systems that are important to society under conditions where the system itself is constantly changing.

Conservationist's outlook has a clear and remarkable difference over stewardship in the sense that conservationists do not approve change in the ecosystem or allow the ecosystem to leverage the services to surrounding stakeholders. The stewardship proponents, on the other hand, dovetail anthropogenic needs as well as functions with the ecosystem services so that all kind of functional changes can well be accommodate by the resilience generated there with. Managing resources to foster resilience —to respond to and shape change in ways that sustain and develop the same fundamental function, structure, identity, and feedbacks seems crucial to the future of humanity and the Earth System. Resilience based ecosystem stewardship is a fundamental shift from steady stateresource management, which attempted to reduce variability and prevent change, rather than to respond to and shape change in ways that benefit society (Table 3.1).

We emphasize resilience, a concept that embraces change as a basic feature of theway the world works and develops, and therefore is especially appropriate at times when changes are a prominent feature of the system. We address ecosystems that provide a suite of ecosystem services rather than a single resource such as fish or trees. We focus on stewardship, which recognizes manager as an integral component of the system that they manage. Stewardship also implies a sense of responsibility for the state of the system of which we are a part (Leopold 1949). The challenge is to anticipate change and shape it for sustainability in a manner that does not lead to loss of future options (Folkeet al. 2003). Ecosystem stewardship recognizes that society's use of resources must be compatible with the capacity of ecosystems to provide services, which, in turn, is constrained by the life support system of the planet (Fig. 3.1).

Table 3.1 Contrasts between steady state resource management, ecosystem management, and resilience based ecosystem stewardship

Steady state resource management	Ecosystem management	Resilience based ecosystem Stewardship
Reference state: historic condition	Historic condition	Trajectory of change
Manage for a single resource or species	Manage for multiple ecosys-tem services	Manage for fundamental social–ecological properties
Single equilibrium state whose properties can be sustained	Multiple potential states	Multiple potential
Reduce variability	Accept historical range of vari-ability	Foster variability and diversity
Prevent natural disturbances	Accept natural disturbances	Foster disturbances that sustain social–ecological properties
People use ecosystems	People are part of the social–ecological system	People have responsibility to sustain future options
Managers define the primary use of the managed system	Multiple stakeholders work with managers to define goals	Multiple stakeholders work with managers to define goals
Maximize sustained yield and economic efficiency	Manage for multiple uses despite reduced efficiency	Maximize flexibility of future options
Management structure protects current management goals	Management goals respond to changing human values	Management responds to and shapes human values

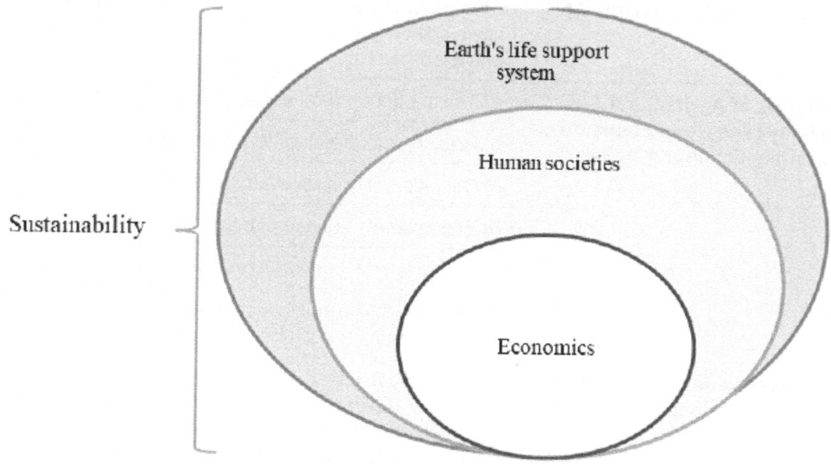

Figure 3.2 Social–ecological sustainability requires that society's economy and other human activities not exceed the capacity of ecosystems to provide services, which, in turn, is constrained by the planet's life support system. Redrawn from Fischer et al. (2007).

This figure delineated above presents a framework for understanding and managing resources in a world where persistent directional changes are becoming more pronounced.

Stewardship and Sustainability Issues

Managing for sustainability requires attention to changing typical of complex adaptive systems. Now we must place sustainability in the context of the directional changes in factors that govern the properties of most social–ecological systems. Three broad categories of outcome are possible: (1) persistence of the fundamental properties of the current system through adaptation, (2) transformation of the system to a fundamentally different, potentially more desirable state, or (3) passive changes (often degradation to a less favorable state) of the system as a result of failure of the system to adapt or transform. Intermediate outcomes are also possible, if some components (e.g., ecological subsystems, institutions, or social units) of the system persist, others transform, and others degrade (Turner et al. 2003). Sustainability implies the persistence of the fundamental properties of the system or of active transformation through deliberate substitution of different forms of capital to meet society's needs in new ways. In contrast, degradation implies the loss of inclusive wealth and therefore the potential toachieve sustainability.

Sustainable agriculture considered from two point of view	
1. Environmental Conservation	
2. Non stagnant productivity	Food grain availability and trend in agro ecosystem consisting of irrigated area, coastal belt, rainfall area, hill and mountain areas.

Key Element of Sustainable Agriculture	Zero tillage crop, waste management, organic matter recycling, soil enrichment, IPM, use of information technology, precision farming.
Key Actors of Agricultural Sustainability (The Round table on environment economy 1994)	a) The impact on rural economy system within its capacity.
	b) Present and projected environment.
	c) Depletion of non renewable resources.
	d) Equitable distribution of benefits
	e) Diversity and innovated resource efficiency and
	f) Participatory productive system and area of information and education for sustainable agriculture.
Parameters to define operational aspect of sustainability (Swaminathan 1991)	Technological appropriability, technical feasibility, Economic viability, environmental soundness, temporal stability, resource use efficiency, local adaptability, social acceptability, social sustainability, political tactness, Administrative Manageability, Cultural desirability, Renewability, Equity and Productivity.

How can we manage the dynamics of change to improve the chances for persistence or transformation? Four general approaches have been identified as ways to foster sustainability under conditions of directional change: (1) reduced vulnerability, (2) enhanced adaptive capacity, (3) increased resilience, and (4) enhanced transformability. Each of these approaches emphasizes a different set of processes by which sustainability is fostered.

Vulnerability addresses the nature of stresses that cause change, the sensitivity of the system to these changes, and the adaptive capacity to adjust to change.

Adaptive capacity addresses the capacity of actors or groups of actors to adjust so as to minimize the negative impacts of changes.

Resilience incorporates adaptive capacity but also entails additional system level attributes of social–ecological systems that provide flexibility to adjust to change.

Transformability addresses active steps that might be taken to change the system to a different, potentially more desirable state.

Although anthropologists, ecologists, and geographers developed these approaches somewhat independently (Janssen et al. 2006), they are becoming increasingly integrated (Berkes et al. 2003, Turner et al. 2003, Young et al. 2006). This integration of ideas provides policy makers and managers with an increasingly sophisticated and flexible tool kit to address the challenges of sustainability in a directionally changing world. Resilience based ecosystem stewardship accommodates allthese factors interactively performing in a common system so that ecosystem functioncan go sustainable and keep disposing of critical ecological services to mitigate social needs.

Table 3.2 explains the conceptual framework that is linking human adaptive capacity, vulnerability, resilience, and transformability. The system (e.g., household, community, nation, etc.) responds to a suite of interacting drivers (stresses, events, shocks) to produce one of three potential outcomes: (1) persistence of the existing system through resilience; (2) actively navigated transformation to a new, potentially more beneficial state through transformability; or (3) unintended transformation to a new state (often degraded) due to vulnerability and the failure toadapt or transform. These three outcomes are not mutually exclusive, because some components (e.g., ecological subsystems, institutions, or social units) of the system may persist, others transform, and others degrade. The sensitivity of the system to perturbations depends on its exposure (intensity, frequency, and duration) to each perturbation, the interactions among distinct perturbations, and critical properties of the system. The system response to the resulting impacts depends on its adaptive capacity (i.e., its capacity to learn, cope, innovate, and adapt). Adaptive capacity, in turn, depends on the amount and diversity of social, economic, physical, and natural capital and on the social networks, institutions, and entitlements that influence how this capital is distributed and used. System response also depends on effectiveness of cross scale linkages to changes occurring at other temporal and spatial scales. Those components of the system characterized by strong stabilizing feedbacks and adaptive capacity are likely to be resilient and persist. Alternatively, if the existing conditions are viewed as untenable, a high adaptive capacity can contribute to actively navigated transformation, the capacity to change to a new, potentially more beneficial state of the system or subsystem. If adaptive capacity of some components is insufficient to cope with the impacts of stresses, they are vulnerable to unintended transformation to a new state that often reflects degradation in conditions.

Table 3.2 Assumptions of frameworks addressing long term human well being

	Assumed change in	Nature of mechanisms	Other approaches
Framework	exogenous controls	emphasized	often incorporated
Vulnerability	Known	System exposure and sensitivity to drivers; equity	Adaptive capacity, resilience
Adaptive capacity	Known or unknown	Learning and innovation	None
Resilience	Known or unknown	Within system feedbacks and adaptive governance	Adaptive capacity, transformability
Transformability	Directional	Learn from crisis	Adaptive capacity, resilience

VULNERABILITY

To organize conservational or stewardship function, an estimation of vulnerability is essential. If an ecosystem is not kept deterrent of vulnerability, thecomponents of that very system will go more fragile against both the internal or external stresses. Vulnerability is the degree to which a system is likely to experience harm due to exposure to a specified hazard or stress (Turner et al. 2003, Adger 2006). Vulnerability theory is rooted in socioeconomic studies of impacts of events (e.g., floods or wars) or stresses (e.g., chronic food insecurity) on social systems but has been broadened to address responses of entire social–ecological systems. Vulnerability analysis deliberatelyaddresses human values such as equity and well being. Vulnerability to a given stress can be reduced by (1) reducing exposure to the stress (mitigation); (2) reducing sensitivity of the system to stress by sustaining natural capital and the components of well being, especially for the disadvantaged; and/or (3) increasing adaptive capacity and resilience (see below) to cope with stress (Table 1.3; Turner et al. 2003).

Table 3.3 Principal sustainability approaches and mechanisms. Adapted from Levin (1999), Folke et al. (2003), Turner et al. (2003), Chapin et al. (2006a), Walker et al. (2006)

Vulnerability
Reduce exposure to hazards or stresses
Reduce sensitivity to stresses
Sustain natural capital
Maintain components of well being
Pay particular attention to vulnerability of the disadvantaged
Enhance adaptive capacity and resilience (see below)
Adaptive capacity
Foster biological, economic, and cultural diversity
Foster social learning
Experiment and innovate to test understanding
Select, communicate, and implement appropriate solutions
Resilience
Enhance adaptive capacity (see above)
Sustain legacies that provide seeds for renewal Foster a balance between stabilizing feedbacks and creative renewal
Adapt governance to changing conditions
Transformability
Enhance diversity, adaptation, and resilience
Identify potential future options and pathways to get there
Enhance capacity to learn from crisis
Create and navigate thresholds for transformation

Exposure to a stress can be reduced by minimizing its intensity, frequency, duration, or extent. Prevention of pollution or banning of toxic pesticides, for example, reduces the vulnerability of people who would otherwise be exposed to these hazards. Mitigation (reduced exposure) is especially challenging when the stress is the cumulativeeffect of processes occurring at scales that are larger than the system being managed. Anthropogenic contributions to climate warming through the burning of fossil fuels, for example, is globally dispersed, so it cannot be reversed by actions taken solely by those regions that experience greatest impacts of climatic change (McCarthy et al. 2005).

Sensitivity to a stress can be reduced in at least three ways: (1) sustaining the slow ecological variables that determine natural capital; (2) maintaining key components of well being; and (3) paying particular attention to the needs of the disadvantaged segments of society, who are generally most vulnerable. The poor or disadvantaged, for example, are especially vulnerable to food shortages or economic downturns, and people living in floodplains or the wild land–urban interface are especially vulnerable to floodingor wildfire, respectively. An understanding of the causes of differential vulnerability can lead to strategies for targeted interventions to reduce overall vulnerability of the social– ecological system.

ADAPTIVE CAPACITY

Adaptive capacity (or adaptability) is the capacity of actors, both individuals and groups, to respond to, create, and shape variability and change in the state of the system (Folke et al. 2003, Walker et al. 2004, Adger et al. 2005). Although the actors in social– ecological systems include all organisms, we focus particularly on people in addressing the role of adaptive capacity in social–ecological change, because human actors basetheir actions not only on their past experience but also on their capacity to plan for the future (reflexive action). This contrasts with evolution, which shapes the properties of organisms based entirely on their genetic responses to past events. Evolution has no forward looking component. Adaptive capacity depends on (1) biological, economic, andcultural diversity that provides the building blocks for adjusting to change; (2) the capacity of individuals and groups to learn how their system works and how and why itis changing; (3) experimentation and innovation to test that understanding; and (4) capacity to govern effectively by selecting, communicating, and implementing appropriate solutions.

The role of human capital in adaptive capacity is especially important. It is much more than formal education. It depends on an understanding of how the system responds to change, which often comes from experience and local knowledge of past responses to extreme events or stresses. As the world changes, and new hazards and stresses emerge, this understanding may be insufficient. Willingness to innovate and experiment to test what has been learned and to explore new approaches is crucial to adaptive capacity. Social capital through networking to select, communicate, and implement potential solutions is another key component of adaptive capacity. In the context of sustainability, adaptive capacity represents the capacity of a social–ecological system to make appropriate substitutions among forms of capital to maintain or enhance inclusive wealth. In this way the system retains the potential for future generations to meet their needs.

RESILIENCE

Resilience is the capacity of a social–ecological system to absorb a spectrum of shocks or reorganization in a new context (Holling 1973, Gunderson and Holling 2002, Walker et al. 2004, Folke 2006). The unique contribution of resilience theory is the recognition and identification of several possible system properties that foster renewal and reorganization after perturbations (Holling 1973). Resilience depends on (1) adaptivecapacity (see above); (2) biophysical and social legacies that contribute to diversity and provide proven pathways for rebuilding; (3) the capacity of people to plan for the long term within the context of uncertainty and change; (4) a balance between stabilizing feedbacks that buffer the system against stresses and disturbance and innovation that creates opportunities for change; and (5) the capacity to adjust governance structures to meet changing needs (Holling and Gunderson 2002, Folke et al. 2003, Walker et al. 2006;).

Loss of resilience pushes a system closer to its limits. When resilience has been eroded, a disturbance, like a disease, storm, or stock market fluctuation, that previously shook and revitalized the resilient system, might now push the fragile system over a threshold into an alternative state (a regime shift) with a new trajectory of change. Such system changes radically alter the flow of ecosystem services and associated livelihoods and well being of people and societies. Clearly, resilience is an essential feature of resource stewardship under conditions of uncertainty and change, so this approach to resource management is even more important today than it has been in the past.

Fostering small scale variability and change logically contribute to resilience because it maintains within the system those components that are well adapted to each phase of the adaptive cycle—ranging from the renewal to the conservation phase. This reduces the likelihood that the inevitable disturbances will have catastrophic effect.Conversely, preventing small scale disturbances such as insect outbreaks or fires tends toeliminate disturbance adapted components, thereby reducing the capacity of the systemto cope with disturbance, Suite of early succession species that are well adapted to post disturbance conditions, but may be uncommon in the mature forest. Similarly, the stories and memories of elders and the written history of past events often provide insight into ways in which people coped with past crises as well as ideas for future options that mightnot otherwise be considered. This often occurs by drawing on **social memory, the social legacies** of knowing how to do things under different circumstances. (Chapin et al, 2009)

Ecosystem Stewardship

An ecosystem consists of organisms (plants, microbes and animals including people) and the physical components (atmosphere, soil, water, etc.) with which they interact. All ecosystems are influenced, to a greater or lesser degree by social processes (i.e., are social ecosystem), although ecosystem studied tend to focus on biologicalinteractions.

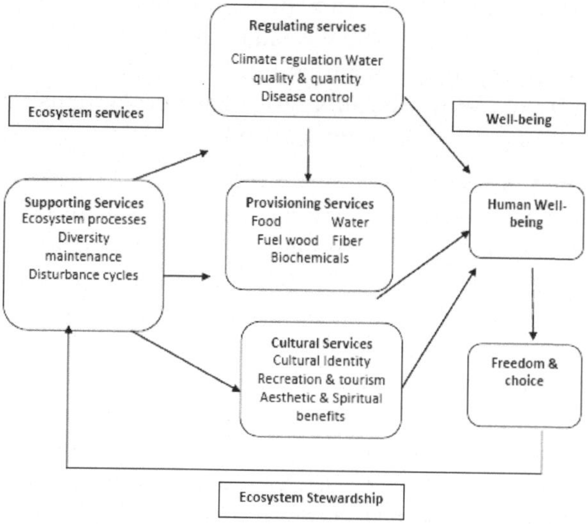

Fig 3.3 Linkages among ecosystem services, well being of society, and ecosystem stewardship, a framework developed by the Millennium Ecosystem Assessment (MEA, 2005c).

Supporting Services: Sustaining Ecosystem Functioning

Supporting services are the fundamental ecological processes that control the structure and functioning of ecosystems. Because of the fundamental dependence of all ecosystem services on supporting services integrity of these services generally sustains many services that are valued more directly by society.

Maintenance of Soil Resources:

Soil and sediments are key slow variables that regulate ecosystem processes by providing resources required by organism. The controls over the formation, degradation and resource supplying potential of soils and sediments are therefore central to sound ecosystem management and to sustaining the natural capital on which society depends (Birkeland 1999, Chaping et.al 2002). The quantity of soil in an ecosystem depends largely on the balance between inputs from weathering (the breakdown of rocks to form soil) or deposition or losses from erosion. In addition, organisms, especially plants, add organic matter to soils through death of tissues and individuals, which is offset by losses through decomposition. If an ecosystem were at steady state, i.e., when inputs approximately equal outputs, the quantity of soil would remain relatively constant, providing a stable capacity to supply vegetation with water and nutrients. Natural imbalances between inputs and outputs lead to deeper soils in floodplains and at the base of hills than on hilltops. Accelerated soil erosion is one of the most serious causes of global declines in ecosystem services and resilience. Soil erosion from land represents a sediment input to lakes and estuaries. At a global scale the increased sediment input to oceans from accelerated erosion is partially offset by the increased sediment capture by reservoirs. Therefore lakes, including reservoirs, and estuaries are the aquatic ecosystems most strongly affected by terrestrial erosion. Especially in agricultural areas, these sediments represent a large influx of nutrients (eutrophication) to aquatic ecosystems that can be just as problematic as loss of productive potential on land.

Roberts defined **Stewardship** or land ethic as a set of values which engenders an appreciation of and respect for the land as the basis of our prosperity and quality of life , and explained that adopting land ethic was a new concept for most Australians (1992:17). For Vanclay (1992:97), **Stewardship** refers to the notion that farmers are stewards of the land and that farming is a way of life that places implicit responsibilityon farmers to look after the land for future generation. The **Stewardship** concept recognizes that farmers may have to make uneconomical decisions in order to protect theland. The analysis of survey by Vanclay (1986) suggested that Landholders are more concerned by economic impacts of land degradation. In this survey property size was part of the explanation for the absence of the predicted linkage between land care and **Stewardship**. Landcare respondents were significantly more likely to operate larger farms. However smaller property owners had significantly higher scores on the stewardship scale. Survey result confirmed that most of the farmer have a strong stewardship ethic and indicated that majority of landholders were concerned about the environmental impact of land degradation. A stronger mix of policy instrument is neededto support land care. e.g. Wetland management is an excellent example of the mix of policy option needed to support landcare.

Soil stewardship acts as a flagship case for the intrinsic value of natural resources and provides a valuable example for education of links between land use and health (Aidan M. keith et al. 2016). A synergy between different separate approaches would benefit from a consolidation of relevant knowledge from the large body of existing literature, following which research gaps, or areas lacking in studies, could also be identified systematically. Studies examining the effects of land management and its change on arrange of ecosystem services are becoming familiar, but those making links to animal, human and ecosystem health are less abundant (e.g. Rhodes et al.,2013). Aidan et al. highlighted four broad areas for consideration:

1. Relationship between soil stewardship practices and health metrics at farm and landscape scales.
2. Mediation of exposure to chemical contaminants by soil stewardship practices; contamination through the food chain (e.g. uptake by crops of veterinary pharmaceuticals via manures) and transport to air/water.
3. Mediation of the dynamics of pathogenic organisms important to crop, animal and human health by soil stewardship practices, including landscape configuration.
4. Correlations and trade-offs between ecosystem services and health metrics at landscape and regional scales.

Large scale environmental and soil surveys (e.g. Land Use/Land Cover Area Frame Survey [LUCAS], GB Countryside Survey) and research site networks (e.g. Long Term Ecological Research network), which generally have an array of collocated measures, could be exploited to help investigate such relationships. The call for soil security risk assessments by McBratney et al. (2014) could also be adapted to incorporate risks to animal, human and ecosystem health, thus encapsulating potential knock on effects to those dependent on these soils.

WATER CYCLING

Water is the soil resource that is used in largest quantities by plants and which most frequently limits the productivity of terrestrial ecosystems. Water enters ecosystems as precipitation. Two of the major pathways of water loss are transpiration, the "green water" that support terrestrial production, and runoff, which replenishes groundwater and aquatic ecosystems, the "blue water" sources that are often tapped by people for domestic and industrial uses, irrigation , and hydropower. Consequently, there are inherent tradeoffs among the ecosystem services provided by water cycling, and some of the biggest challenges in water management result from these tradeoffs.

Climatic controls over water inputs in precipitation place an ultimate constrainton quantities of water cycled through ecosystems. Within this constraint the partitioning of water between transpiration and runoff depends on (1) the degree of compaction ofsoil surface, which influences water infiltration into the soil, (2) Soil water holding capacity, which depends on the quantity of soil and its particle size distribution, and (3) the capacity of vegetation to transpire water (Rockstrom et al. 1990).

Vegetation fosters infiltration and storage through several mechanisms. The plant canopy and litter reduce compaction by raindrops that otherwise tend to seal soil pores. In addition roots and soil animals associated with vegetation create channels for water movement through the soil profile. By facilitating water infiltration (due to reduced compaction), water holding capacity (due to production of soil organic matter), and reduced soil erosion (due to reduced overland flow), vegetation generates stabilizing feedbacks that sustain the productive potential of soils and reduced ecosystem vulnerability to drought. Human activities often disrupt these stabilizing feedbacks,thereby reducing resilience. Plowing reduces soil organic content sustainability often by 50% over a few years thereby reducing soil water holding capacity and the capacity of soils to support crop growth with natural rainfall. Alternative agriculture practices that conserve or rebuild soil organic content (e.g., no till agriculture) or reduce compactionby livestock or equipment under wet conditions therefore increase the capacity of soils tosupply water to crops or other vegetation.

AGRICULTURAL WATER STEWARDSHIP BUILDS AGRICULTURAL AND ECOLOGICAL RESILIENCE

Agricultural water stewardship can be defined as the use of water in a manner that optimizes agricultural water use while addressing the co benefits of water for food production, the environment, and human health. Agricultural water stewardship is premised on the notion that water management decisions cannot be made independent of ecological, social, and economic contexts. It takes into account local biogeographical considerations such as soil type, soil ecology, topography, agronomic inputs, and terrestrial and aquatic ecology. Agricultural water stewardship emphasizes whole farm approaches that acknowledge the role of agriculture in the local watershed and community. Effective stewardship results in the provision of multiple types of ecosystemservices—the human benefits garnered from well functioning ecosystems, including agro ecosystems. In summary, the local and regional context in which a farm operates will determine the best water management approaches to maximize water use efficiency, minimize adverse environmental impacts, and balance trade offs for society as a whole. Agricultural water stewardship takes into account a broad range of variables in identifying balanced solutions. Importantly, it embraces the full spectrum of solutions, from cutting edge technology to the cultural and agro ecological practices that enhance the ecological integration of farming and lead to improved water quality and supply management.

BENEFICIAL MANAGEMENT PRACTICES

Many practices that could enhance performance on a wide range of water stewardship indicators are already known, but their application should be expanded. Examples of beneficial practices include:

On farm stewardship practices

- Use of irrigation systems that are designed for optimum "distribution uniformity," with appropriate management to also achieve a high efficiency for each irrigation and location
- Improvement of relative timing of water applications and chemical and nutrient inputs to minimize aquifer pollution
- Tail water ponds and return systems
- Vegetated filter strips
- Efforts to reinstate natural flows and hydrograph cycles
- Restoration of riparian corridors and other waterways for multiple ecosystem services (e.g., infiltration, soil and water quality, wildlife habitat, etc.)
- Soil management regimes to enhance infiltration and retention of water in soils, including cover crops, amendments, minimum tillage

- Soil moisture and crop water status monitoring
- Wind turbines for frost control
- Dry farming

Supply management

- Off stream storage ponds for irrigation and frost protection
- Managed groundwater recharge using surplus irrigation water when chemical ornutrient (e.g., nitrate) runoff is low
- Wetlands for improving water quality and wildlife habitat
- Recycling of municipal water for agricultural use, with proper attention to waterquality
- Measures to better match quality of water with intended use, including drinkingwater for rural communities
- Modification of conveyance capabilities to move flows to recharge areas

(This list is intended to illustrate the range of possible practices and is not an exhaustiveor prioritized list)

Fresh Water Facts

2. Nearly 80 per cent of the world's population is facing threats to water security
2. More than 64 per cent of the world's wetlands have been lost in the 20th century
3. The Freshwater Living Planet Index, which measures trends in vertebrate species populations, shows a decline of 76 per cent between 1970 and 2010
4. Hydropower produces more than one fifth of the world's electricity

THE WATER STEWARDSHIP APPROACH

As more river basins face closure from over allocation of water resources, thereis an opportunity to support basin level approaches and to better understand water management choices and their implications. For business, a broader question is what interventions can deliver the changes to catchment water management necessary toaddress their water concerns. To address the physical risk of water availability, it may be possible in some cases to make local changes to enhance hydrology to meet operational water requirements. In many instances, however, multiple water users are drawing on the same water resources and it is not possible to exclude other users from access to water saved by any single facility or intervention. Furthermore, this still does not address the potential reputational or regulatory risks generated when interventions inadvertentlyincrease consumptive use and reduce downstream flows. In most cases it is unlikely that isolated field level interventions by individual businesses can be achieved at the necessary scale or with sufficient continuity to meaningfully influence water flows. Supply chain risks cannot fully be addressed by solely looking at farm level water management. Scaling up requires collective action with other water users and engagement with water governance institutions to achieve optimized, sustainable allocations of water and maintenance of environmental flows and ecosystem services10. To address fundamental water risks, direct business interventions in agricultural water management need to be part of a broader strategy to strengthen water governance and improve river basin planning. Interventions must explicitly recognize other actors working within catchments, and genuinely and materially address shared water issues that drive corporate water risks. This requires speaking with, learning from and collaborating with water management agencies and regulatory authorities who have to manage water at multiple scales. It should also prompt a review of what crops are grown and for what purpose, as well as alternative uses of water – for energy, ecosystem services or drinking water security. Collective action must ensure that water users jointly determine optimal water resource use, rather than allowing powerful commercial interests to dominate water management. Bottom up collective action should help bring coherence to a currently scattered set of small scale interventions and match these to large scale infrastructure and systems where relevant. While field level interventionsmay be insufficient to truly mitigate a full range of water related business risk, more comprehensive and robust monitoring is needed to understand whether they create real water savings or affect catchment hydrology. If the evidence shows positive impacts, beneficial projects can be scaled up by those with a governance mandate for water management.

As the discipline of water stewardship matures and moves from high level discussion and concepts to business action, there is a parallel mobilization of practitioners and academics providing robust scrutiny of both the intended and unintended outcomes. It is critical, therefore, that businesses engaged in water stewardship fully understand the hydrological, sociopolitical consequences of their water management interventions. A failure to appreciate these complexities can lead to overly simplistic, supply side, techno centric solutions that will not mitigate risks, nor deliver sustainable water management at the basin scale. Some key considerations for businessesengaging in agricultural water management are:

i. Improve site level transparency of water information: The lack of transparency and data impedes verifying if real water savings has occurred.

ii. Improve availability of basin level water stocks and flows: Information on surface and groundwater levels and fluxes is critical to creating effective solutions.

iii. Focus on shared water challenges, not efficiency solutions: Companies must understand whether interventions effectively address water challenges and reduce water risk, or simply transfer the challenges and risks to others downstream.

iv. Understand the scales of hydrology, from local to catchment or basin: Failing to understand how water flows through a system can undermine the effectiveness of the intervention.

v. Understand water's role in the economy: Evaluate trade offs between the use of water for food, energy, ecosystems or greater water security.

vi. Be clear on definitions: In particular, pay attention to how terms such as "water efficiency," "water productivity," and "saved water" are used.

vii. Understand allocations: Relate your water management activities to the basin and ask yourself how you are supporting water management in policy and in practice.

viii. Seek solutions at the field and policy level in parallel: Only undertake field scale interventions in water management if there is a parallel process of policy engagement to implement an effective framework of water allocations.

ix. Understand there is intense competition for "saved" resources; don't think you are banking them for your exclusive future use. Initiate a dialogue with parties interested in these savings. (Water Stewardship in Agriculture, WWF Brief 2015, Together in partnership with IUCN)

MAINTENANCE OF BIOLOGICAL DIVERSITY

The known effect of biodiversity on ecosystem functioning relate more strongly to traits that govern the effect of species on ecosystem processes and trait that govern the response of species to environmental variation. These known effects of biodiversity change (both loss of key species or invasion of species with large impacts)can befostered by maintaining species that span the spectrum of effect diversity and response diversity present in the system (Elmqvist et al.2003, Suding et al. 2008).

The functional redundancy associated with species diversity provides insurance against more drastic changes in environment, such as those that may occur in the eventof human mismanagement of ecosystems or change in climate. Radical changes inenvironment are unlikely to eliminate all species of a given functional type in a diverse ecosystem, allowing the surviving species to increase in abundance and maintain functions that might otherwise be seriously compromised.

Table: 3.4 Examples of biodiversity effects on ecosystem services. (Chapin et al. modified it from Diaz et al. 2006)

Ecosystem Service	Diversity component and mechanism
1. Production by societally important plants	Functional composition : (a) Fast growing species produce more biomass; (b) species differ in timing and spatial pattern of resource use(complementarily allows more resource to be used) Species number: large species pool is more likely to contain productive species.

Ecosystem Service	Diversity component and mechanism
2. Stability of crop production	Genetic diversity: buffers production against losses to pests and environmental variability. Species number: Cultivation of multiple species in the same plot maintains high production over a broader range of conditions. Functional composition: species differ in their response to environment and disturbance, stabilizing production.
3. Maintenance of soil resources	Functional composition: (a) fast growing species enhance soil fertility; (b) Dense root systems prevent soil erosion.
4. Regulation of water quantity and quality	Landscape diversity: Intact riparian corridors reduce erosion. Functional composition: Fast growing plants have high transpiration rates, reducing stream flow.
5. Pollination for food production and species survival	Functional composition: loss of specialized pollinators reduces fruit set and diversity of plant that reproduce successfully. Species number: Loss of pollinator species reduces the diversity of plants that successfully reproduce (genetic impoverishment) Landscape diversity: Large, well connected landscape units enable pollinators to facilitate gene flow among habitat patches.
6. Resistance to invasive species with negative ecological/cultural effects	Functional composition: Some competitive species resist the invasion of exotic species. Landscape structure: Roads can serve as corridors for spread of invasive species; natural habitat patches can resist spread. Species number: Species rich communities are likely to have less unused resources and more competitive species to resist invaders.
7. Pest and disease control	Genetic diversity or species control: Reduces density of suitable hosts for specialized pests and diseases. Landscape diversity: Provides habitat for natural enemies of pests.
8. Biophysical climate regulation	Functional composition: Determines water and energy exchange, thus influencing local air temperature and circulation patterns Landscape structure: Influences convective movement of air masses and therefore local temperature and precipitation.
9. Climate regulation by carbon sequestrations	Landscape structure: Fragmented landscapes have greater edge to edge ration; edges have greater carbon loss. Functional composition: Small, short lived plants store less carbon. Species number: High species number reduces pest outbreaks that cause carbon loss.
10. Protection against natural hazards (e.g., floods, hurricanes, fires)	Landscape structure: Influences disturbances spread and/or protection against natural hazards. Functional composition: (a) extensive root systems prevent erosion and uprooting; (b) deciduous species are less flammable than evergreens

OPERATIONAL CONCEPT OF FARM STEWARDSHIP

The "modernization of agriculture" in the decades of 50s and 60s had really been a response to political economy of hunger, poverty and silence; which in turn left behind the green revolution and array of residual hazards. These residual hazards are both biophysical and socio ecological. The reminiscence of green revolution have been traced out in the form of contaminated ground water, soil loaded with problems of salinity, and of course, the gross erosion of biodiversity. While the contamination of water and erosion of soil have made farm production process a more complex one, the erosion of biodiversity has offered Indian agriculture a fragile and narrow gene base." (Quoted from 'The Silent Spring' Richard Carlson) of green revolution have been traced out in the form of contaminated ground water , soil loaded with problems of salinity, and of course , the gross erosion of biodiversity. While the contamination of water and erosion of soil have made farm production process a more complex one, the erosion of biodiversity has offered Indian agriculture a fragile and narrow gene base." (Quoted from 'The Silent Spring'; Richard Carlson)

So, what's the role of stewardship in farming or agricultural production process at large? The Farm Stewardship is both the process and approach towards making farm production process more resilient , elastic and productive in a given ecological set up to make provisions for ecological deliverables there in and therewith in supporting community people along with other life forms.

Hence, this may be treated as the operational definition for farm stewardship; else it's a conceptual and empirical improvisation over classical conservation strategy. The basic component of stewardship for Indian farms can be derived in the following ways:

I. Base line information on input application, management practices and production status for performing farm and their interrelationship with ambience institutions.

II. Optimization of cost intensive chemical inputs in favor of organic input driven production process.

III. To estimate the sub ecological interactions and relations of the farms so that integrated and resilient farm production process can be ushered.

IV. The social compatibility, economic profitability and ecological suitability need to converge on a common agendum of deliverables with design and impact.

The application of the concept in the present empirical research has been proposed to follow both the inductive and deductive approaches. The approaches are as follows:

I. The study on connectivity between conservation of soil, water, biodiversity in terms of interdependent and interaction these will follow application of both structure schedule and participatory approach

II. The study on connectivity between stewardship of soil, water, biodiversity in terms of interdependent and interaction. These will follow application of both structure schedule and participatory approach.

III. To derive the possibility of applying farm stewardship approaches through necessary transformation, up gradation and acculturation of conservation strategies which are being followed mostly by the practicing farmers.

IV. Identification of conceptual motivational and institutional constraints acting as a barrier to this desired transformation.

V. Through participatory designing and analysis, we can identify the people –ready status for ushering farm stewardship in terms of economic viability, social compatibility and ecological resilience.

ORIGIN AND SOCIALIZATION OF THE CONCEPT

As a concept stewardship is western in origin and traditionally followed by a group of professional there to protect water, soil and forest from abrupt and aggressive depletion by anthropogenic activities. The present research envisages exploring the possibilities of functional stewardship as applicable to Indian Agriculture. The extrapolation of conservation approaches can therefore be applied to configure stewardship functions. Steward—" A person who manages another's property or financial affairs; one who administers anything as an agent of another or others." (The random house dictionary of the english language-1868)

A landowner's claim to stewardship of his or her land for a particular purpose implies a tremendous responsibility. The steward cannot treat the land as only his or her own property, but must manage it as if acting as the agent for others, as if the land belonged to others. In the ecological context, therefore, land stewardship must mean that the landowner manages the land with the purpose of minimizing the ecological footprint in mind. The landowner claiming to be engaged in sound ecological stewardship must demonstrate not only that his or her land is ecologically vital, but that management of theland does not place anyone else's land in ecological jeopardy. That is the test of land stewardship in the ecological sense. That is the test to which we must put any claim by farmers that they are acting as good ecological stewards of their land.

This proclamation is the mantra of the "first stewards of the land" rhetoric of agricultural policy. The argument is that because farmers "depend" on their land, becausethey "know" and "love" their land, they are environmentally benign or, even better, a positive environmental force. All we need to do is let farmers do the thing that comes naturally to them, that flows from their love for and knowledge of the land, and everything will be alright.

The Stewardship function is multifunctional although having a start from land based production system operating along and across the time, cost and resource continuum. In Indian agriculture the worst affected component of farm ecology are soil, water and biodiversity. The ecological deliverables in the form of product or by product for example paddy and straw are not supported by ample market prices, good enough to say that agriculture is turning more non remunerative than any time before. For farmer every cost is certain or increasing, and every return is uncertain or declining. The stress from market vagaries has come out to be deleterious for both social ecology as well as farm ecology.

A desperation among the farmers has been reflected to go for shorter duration of crop , moderate to higher level of productivity just to catch up market buoyancy, whichin turn has provocated them go for high intensity application of agri chemicals, fertilizers and uncalled for management of resources. The ephemeral nature of management has damaged the prospect of sustainability and resilience in ecosystem. Trends are also becoming conspicuous where in higher agricultural production is inviting more agony to the farmers due to market plunge and livelihood losses.

So, just soil and water conservation can't be a panacea for ecological restoration, rather we have to go beyond for a qualitative transformation of classical conservation approaches to more belligerence stewardship intervention. For example, Raising a short duration leafy vegetables out of residual moisture and minimal input application can add more meaning to conservation of water , better to term it a responsible stewardship. Another example : Excavating a seepage tank may help conservation of runoff water but harvesting some quintals of quick growing seasonal fish out of this water conserved may add not only to an economic value to this conservation approaches , the other way round its better be branded as a socially orchestrated stewardship function.

STEWARDSHIP: ECOLOGICAL GOODS AND SERVICES

The stewardship function is all about extending ecological functions to generate ecological goods and services respectively. While explicit recognition of ecosystem services is a relatively new phenomenon, the notion that natural ecosystems help to support society can be traced to ancient times. Although the term "ecosystem services" was first used in the late 1960s, it in is only within the last decade that the concept has gained hold in the broader research, policy and natural resource management community (Whitten et. al 2003). A brief history of the ecological and goods and services concept is provided by Mooney and Ehrlich (1997).

Daily notes that ecosystem services are the conditions and processes through which natural ecosystems, and the species that make them up, sustain and fulfill human life. Biophysical systems with ecological integrity maintain biodiversity and the production of ecosystem goods, such as seafood, forage, timber, biomass fuels, natural fiber, and many pharmaceuticals, industrial products, and their precursors. In addition to the production of goods, ecosystem services are the actual life support functions, such as cleansing, recycling, and renewal, and they confer many intangible aesthetic and cultural benefits as well (Daily 1997b). Costanza et al. note that ecosystem functions refer variously to the habitat, biological or system properties or processes of ecosystems. Ecosystem goods (such as food) and services (such as waste assimilation) represent the benefits human populations derive, directly or indirectly, from ecosystem functions (Costanza et al. 1997).

DEFINING SUSTAINABLE AGRICULTURE

According to Schaller, sustainable agriculture, like the concept of sustainability itself, suggests not only the core characteristics of a destination for agriculture but also particular farming practices that could move agriculture towards that destination (with both ends and means). He points out, however, that neither characteristic lends itself to precise definition. "As a destination," he writes, "sustainability is not easily captured in concise definitions. On the other hand, specific farming practices need to be tailored to the particular location and circumstance, continual modification and adaptation." Because of these difficulties, he notes, many people prefer to focus on the technical side of sustainable agriculture, about the different kinds of practices that may conserve soil and water, protect the environment, and provide the farmer with a decent profit. Scientists in particular, Schaller argues, tend to think of the ends of sustainability as "givens" and the role of science to determine how best to achieve them. For that reason and because it is difficult and unwise to issue a top down definition of sustainable farming practices, Schaller suggests that people's beliefs and values play a role in determining what sustainability is and how it can and should be achieved (Schaller 1993).

Schaller notes that current beliefs and values differ markedly. At one extreme there are those who believe that conventional agriculture is basically sound and that fine tuning or modification will suffice. According to Schaller, this view is still widely held by many people and organizations within the traditional agricultural community, such as the U.S. Department of Agriculture, the land grant colleges of agriculture, farm and commodity organizations, and business firms. In contrast are those who feel that a fundamentally different way of thinking about agriculture is required in order to make it sustainable (Schaller 1993).

Another useful framework for understanding the different conceptualizations of sustainable agriculture is provided by Douglass (1984). He notes that "agricultural sustainability" can be defined in different ways and sought through different means. He defines three main approaches. The first group defines agricultural sustainability in economic or production terms, focusing on food sufficiency. A second group defines it in biophysical terms and stewardship. A third group, the "alternative" agriculturalists, are concerned with sustaining the permanent carrying capacity of renewal agricultural resources, but also focus special attention on the effects of different agricultural systems on the social organization and culture of rural life. There is a concern for the values of stewardship, self reliance, humility and holism. They also believe that the agricultural information encoded in the cultural practices of traditional rural societies is as important to the creation of sustainable agricultural systems as the products of science (Douglass 1984).

The rise of this "alternative" group can be traced back to Rachel Carson's 1962 work Silent Spring, a criticism of pesticides, which served as a catalyst for the emergence and growth during the 1970s and 1980s of an agricultural sustainability movement that focused on the social and environmental externalities associated with modern agricultural technologies. Kloppenberg et al. (2000) noted that the success that farm and agrarian activists enjoyed in deconstructing the legitimacy of industrial agriculture was reflected in the National Research Council (1989) report, Alternative Agriculture, which called for making alternative practices more widespread.

Kloppenberg et al. (2000) also noted that although the "alternative" agricultural sustainability movement found its origins in farm environmental issues, globalization in the 1990s has prompted a broader approach to the social analysis of food extending far beyond the farm gate. The failure of many proponents of sustainable agriculture to adequately treat social injustice (class, gender, and especially hunger), as Kloppenberg et al. (2000) pointed out, has engendered considerable criticism. Allen, for example, has argued that, historically, sustainable agriculture has focused on the environmental and there is a need to include human needs and social relations. Working to fulfill basic human needs, she argues, "requires reframing our concept of sustainability to include a social dimension and a concomitant expansion of our approach to sustainable agricultural research." (Allen 1993). Allen calls for the development of "new epistemological and research approaches that integrate the natural processes of ecology combined with the social relations compatible with sustainability. This will involve, for instance, examining not only techniques for reducing soil erosion and water depletion, but also new forms of social organization that alternative agricultural practices presuppose" (Allen, 1993). Study and activism around food issues have generally come now to encompass the larger concerns of social justice and environmental interests in additional to traditional agricultural problematics. Those working for the transformation of the food sector now commonly frame their ambitions not in terms of sustainable agriculture per se, but as the realization of a sustainable food system (Kloppenberg et al. 2000).

SUSTAINABLE AGRICULTURE AS ALTERNATIVE AGRICULTURE

The understanding of sustainable agriculture used in this thesis follows from Douglass' third group, the "alternative" agriculturalists. It is a holistic approach to sustainable agriculture which encompasses the goals of food sufficiency and environmental protection, but which also addresses human needs and social relations. The definition of sustainable agriculture used in this thesis, therefore, extends beyond production to take into consideration the entire food system.

Sustainable agriculture integrates three main goals – environmental health, economic profitability, and, social as well as economic equity. Sustainability, as previously noted, rests on the principle that we must meet the needs of the present without compromising the ability of future generations to meet their own needs. It recognizes that natural resources are finite, acknowledges limits on economic growth, and encourages equity in resource allocation. Therefore, stewardship of both natural and human responsibilities is of prime importance. Stewardship of human resources includes consideration of social responsibilities such as working and living conditions of labourers, the needs of rural communities, and consumer health and safety both in the present and the future. Stewardship of land and natural resources involves maintaining or enhancing this vital resource base for the long term.

Taking into consideration the notions of justice and equity, Douglass defines as unsustainable any structure of agricultural production that directly or indirectly worsens the distribution of opportunities or incomes within producing communities, and judges as unsustainable all structures that fail to improve the distribution of opportunities or incomes by a specific degree within a given period of time. Agriculture will be found to be sustainable, he suggests, "when ways are discovered to meet future demands for foodstuffs without imposing on society real increases in the social costs of production and without causing the distribution of opportunities or incomes to worsen" (Douglass 1984).

According to Pretty, sustainable agriculture is any system of food or fibre production that systematically pursues the following goals: a more thorough incorporation of natural processes; a reduction in the use of off farm, external and non renewable inputs; a more equitable access to productive resources and opportunities, and progress towards more socially just forms of agriculture; a greater productive use of local knowledge and practices, an increase in self reliance among farmers and rural people (Pretty 1995).

Ikerd has argued that sustainable agriculture must use farming systems that conserve resources, protect the environment, produce efficiently, compete commercially, and enhance the quality of life for farmers and society overall (Ikerd 1993).

The sustainable agriculture model relies more on management of the internal resources of the farm (i.e. management skills, knowledge and labour) and maximizes reliance on natural, renewable and on farm inputs, and less on purchased commercial inputs in attempting to reduce negative ecological impacts while maintaining economically viable farms. The goal is to develop efficient, biological systems that do not need high levels of material inputs. The sustainability model implies greater reliance on human resources in terms of the quality and quantity of labour and management, and relatively less reliance on land and capital. Thus, sustainable farming systems may require more farm operators, more farm labourers, and more farm families than do conventional farming systems (Ikerd 1993).

Ecosystem stewardship is an action oriented framework intended to foster the social–ecological sustainability of a rapidly changing planet. Recent developments identify three strategies that make optimal use of current understanding in an environment of inevitable uncertainty and abrupt change: reducing the magnitude of, and exposure and sensitivity to, known stresses; focusing on proactive policies that shape change; and avoiding or escaping unsustainable social–ecological traps. As we discuss here, all social–ecological systems are vulnerable to recent and projected changes but have sources of adaptive capacity and resilience that can sustain ecosystem services and human well being through active ecosystem stewardship.

A CALL FOR ECOSYSTEM STEWARDSHIP

Human actions are having large and accelerating effects on the climate, environment and ecosystems of the Earth, thereby degrading many ecosystem services. This unsustainable trajectory demands a dramatic change in human relationships with the environment and life support system of the planet. Here, we address recent developments in thinking about the sustainable use of ecosystems and resources by society in the context of rapid and frequently abrupt change. Western resource management paradigms have evolved from exploitation, where sustainability is not an important consideration, to steady state resource management aimed at maximum or optimum sustainable yield (MSY or OSY, respectively) and efficient production of a single resource, such as fish or trees, to ecosystem management to sustain a broader suite of ecosystem services. Despite its sustainability goal, management for MSY or OSY tends to overexploit targeted resources because of overly optimistic assumptions about the capacity to sustain productivity, avoid disturbances, regulate harvesters' behavior and anticipate extreme economic or environmental events. Ecosystem management seeks to sustain multiple ecosystem services but often uses, as a reference point, historic conditions that are not achievable in a rapidly changing world. Given the challenges of the sustainable use of ecosystems during rapid change, we advocate a shift to ecosystem stewardship. Its central goal is to sustain the capacity to provide ecosystem services that support human well being under conditions of uncertainty and change. Uncertainty has always characterized social–ecological systems and should therefore not be an impediment to action. Such a paradigm shift entails important tradeoffs, particularly between efficiency and flexibility and between immediate and long term benefits. Ecosystem stewardship integrates three broadly overlapping sustainability approaches: (i) reducing vulnerability to expected changes; (ii) fostering resilience to sustain desirable conditions in the face of perturbations and uncertainty; and (iii) transforming from undesirable trajectories when opportunities emerge. Adaptive capacity contributes to all three sustainability approaches. By building on previous research on vulnerability, adaptation, resilience and transformation, ecosystem stewardship provides a perspective that better equips society to manage a spectrum of challenges that vary in certainty and benefit or threat to society. The need is to identify pragmatic strategies that increase the likelihood of socially beneficial outcomes and reduce the risk of bad outcomes. This approach is explicitly focused on human norms, values and well being and must therefore continually be debated and reassessed by stakeholders. (Chapin et al. 2009)

Natural Resources Stewardship Vision: Although duplicating the natural systems that were present prior to settlement is technically unachievable, it is reasonable to expect that through conscientious intervention a more robust natural ecology can be restored across the city for future generations to enjoy. This is the underlying premise of the natural resources stewardship vision.

NATURAL RESOURCES STEWARDSHIP PHILOSOPHY

The stewardship vision promotes an ecosystem based approach to restoration and management. An ecosystem is essentially where things live and represents an interacting group of physical elements (soils, water, plants, animals, etc.) that inhabit a particular place. All of these elements and their interactions need to be considered in developing goals and plans for management. Ecosystem based management views people as part of the community, and that maintaining a healthy ecosystem is the best way to meet human needs as well as those of other organisms in the community. General goals of this philosophy are to:

- Enhance the health of the ecosystems across the city.
- Enhance the biological diversity of the localized native habitats.
- Provide an appropriate balance between resource preservation, recreational use, and community development.
- Maintain the natural integrity of Lino Lakes.
- To establish partnerships and stakeholder involvement with a variety of agencies and citizens in the community to perpetuate sustainable resources in Lino Lakes.

The stewardship plan focuses on achieving a sustainable landscape quality, which is defined as the point at which Lino Lakes and its partners can indefinitely maintain a certain acceptable level of resource quality within the context of realistic limits – which is contingent upon two primary factors:

- Public understanding of and commitment to natural resource preservation and stewardship programs.
- Undertaking ecological restoration and management programs that are scientifically sound.

Through a well defined stewardship program and a concerted, ongoing effort by Lino Lakes, a certain level of confidence can be gained that the current ecological conditions and trends can be reversed and a more sustainable and higher quality landscape achieved. Note, however, that stewardship programs also need to be flexible due to the changing nature of the living systems addressed by the plan. For these reasons,the plan presented here should be viewed as being neither conclusive nor absolute. It is a starting point in an ongoing process that relies on monitoring and research to provide feedback on program effectiveness.

(Natural Resources Stewardship and Water Resources Management Plan)

Ecoagriculture: Integrating Production and Conservation At A LandscapeScale

The original green revolution, sustainable agriculture and natural resource management (with its primary focus on sustaining the resources underpinning production), or even the ecotechnology approach of Swaminathan (1994) with its focus on the farmer's field, all of these have major elements to contribute. Approaches to biodiversity conservation also need to move beyond the wild biodiversity focus of strictly protected areas and the modest goals of integrated conservation and development projects. Sara J. Scherr and Jeffrey A. McNeely, 2007 argue that ecoagriculture—a fully integrated approach to agriculture, conservation and rural livelihoods, within a landscape or ecosystem context—is needed in many regions. This approach of ecoagriculture can well be integrated to have comprehensive idea about the perception and possibilities of farm stewardship.

Eco agriculture landscapes

Eco agriculture explicitly recognizes the economic and ecological relationships and mutual inter dependence among agriculture, biodiversity and ecosystem services. Eco agriculture landscapes are mosaics of areas in natural/native habitat and are as under agricultural production. Effective eco agriculture systems rely on maximizing the ecological, economic and social synergies among them, and minimizing the conflicts.

The term 'landscape' itself is functionally defined, depending upon the spatial units needed or actually managed by the group of stakeholders working together to achieve biodiversity, production and livelihood goals. Eco agriculture landscapes are land use mosaics with:

- 'natural' areas (with high habitat quality and niches to ensure critical elements for habitat or ecosystem services that cannot be provided in areas underproduction), which are also managed to benefit agricultural livelihoods either through positive synergies with production or other livelihood benefits,
- agricultural production areas (productive, profitable and meeting food security, market and livelihood needs), which are also configured and managed to provide a 'matrix' with benign or positive ecological qualities for wild biodiversity and ecosystem services, and
- institutional mechanisms to coordinate initiatives to achieve production, conservation and livelihood objectives at landscape, farm and community scales, by exploiting synergies and managing trade offs among them. The concept of ecoagriculture further recognizes that agriculture dependent rural communities are critical (and sometimes the principal) stewards of biodiversity and ecosystem services. While protected natural areas are essential in eco agriculture landscapes to ensure critical habitatfor vulnerable species, maintain water sources and provide cultural resource, these resources often may be owned or managed by local communities and farmers.

BIODIVERSITY AND ECOSYSTEM SERVICES IN ECO AGRICULTURE LANDSCAPES

Conservation of biodiversity in eco agriculture landscapes embraces all three elements of agricultural biodiversity defined by the Convention on Biological Diversity: genetic diversity of domesticated crops, animals, fish and trees; diversity of wild species on which agricultural production depends (such as wild pollinators, soil micro organisms and predators of agricultural pests); and diversity of wild species and ecological communities that use agricultural landscapes as their habitat (Convention on Biological Diversity, 2002). Although wild biodiversity and ecosystem services are closely linked, they are not synonymous. A landscape with relatively intact wild biodiversity is likely to provide a full complement of ecosystem services. However, many ecosystem services can also be provided by non native species, or by combinations of native and non native species in heavily managed settings such as permanent farms. The implication is that even where wild biodiversity has been significantly reduced to make way for food and fiber production, high levels of ecosystem services can often still be provided through intentional land management practices. On the other hand, managing an eco agriculture landscape for ecosystem services does not automatically ensure that wild biodiversity will be protected adequately. Thus, wild biodiversity and ecosystem services both require explicit consideration in ecoagriculture systems.

ECOAGRICULTURE APPROACHES

Broadly, ecoagriculture landscapes rely on six basic strategies of resource management, three focused on the agricultural part of the landscape and three on the surrounding matrix.

In production areas, farmers sustainably increase agricultural output and reduce costs in ways that enhance the habitat quality and ecosystem services:

- minimize agricultural wastes and pollution,
- manage resources in ways that conserve water, soils, and wild flora and fauna, and
- use crop, grass and tree combinations to mimic the ecological structure and function of natural habitats. Farmers or other conservation managers protect and expand natural areas in ways that also provide benefits for adjacent farmers and communities:
- minimize or reverse conversion of natural areas,
- protect and expand larger patches of high quality natural habitat, and
- develop effective ecological networks and corridors (McNeely & Scherr, 2003).

The relative area and spatial configuration of agricultural and natural components (and other elements, such as physical infrastructure and human settlements) are key landscape design issues (Forman,1995). The conservation of wild species that are highly sensitive to habitat disturbance —as are some of those most endangered or rare globally—requires large well connected patches of natural habitat. But many wild species, including many that are threatened and endangered, can coexist incompatibly managed agricultural landscapes, even in high yielding systems. Numerous approaches to agriculture, conservation and rural development contribute components, management practices and planning frameworks that can be applied in ecoagriculture landscapes. The outcomes of planning and negotiations among the multiple stakeholders in any particular landscape will take diverse forms depending on the context of local cultures and philosophies of land management.

(i) Real costs are associated with the cross sectoral planning and coordination and technical innovations needed to achieve impacts at a landscape scale. These must be considered in prioritizing private, public and civic ecoagriculture investments. Top priorities would be:

(ii) agricultural landscapes located in or around critical habitat areas for wild species of local, national or international importance (e.g. landscapes in the highly threatened habitats of the Atlantic Forest of Brazil, now dominated by farming),

(iii) degraded agricultural landscapes where restored ecosystem services will be essential to achieve both agricultural and biodiversity benefits (such as the dryland farming and pastoral regions of West Africa),

(iv) agricultural landscapes that must also function to provide critical ecosystem services (such as the densely populated landscapes of Europe and Java), and peri urban agricultural systems, where careful management is required to protect ecological, wildlife and human health. No assessment has been done of the geographical scale and location of such priority areas for ecoagriculture development strategies (as distinct from agriculture or conservation led development), but undertaking such analyses is a critical step to guide policy action.

THE STATE OF ECOAGRICULTURE KNOWLEDGE AND PRACTICE

Little effort has been devoted to explicitly pursuing agricultural development and biodiversity conservation objectives jointly at a landscape scale, so experience is poorly documented and the science is immature and poorly synthesized across disciplines. Removing major barriers to the widespread development of ecoagriculture landscapes requires answering questions like the following.

- How can agricultural production systems contribute to conserving biodiversity while maintaining or increasing productivity?
- How can agricultural and natural areas be jointly managed to produce adequateecosystem services, including wildlife habitat, at a landscape scale?

- How can ecoagriculture approaches become more financially viable for farmers and other stakeholders?
- How can communities, institutions and governments mobilize and develop theinstitutions and policies needed for ecoagriculture landscapes?

The current state of knowledge on these four questions is considered below.

PRODUCTION SYSTEMS THAT SUPPORT BIOLOGICALLY DIVERSE ECOAGRICULTURE LANDSCAPES

Since the 1960s, the 'improved seed–fertilizer–pesticide (irrigation)' paradigm has characterized both industrial agriculture in developed countries and the original green revolution in developing countries. This production model involved short term, plot level production of a small number of crops, generally in monoculture stands (to increase efficiency in use of external inputs and mechanization, maximize the flow of natural resources to harvestable products). Wild flora and fauna were considered direct competitors for resources or harvested products, and thus eliminated, while water was diverted from wetlands and natural habitats for irrigation. More ecologically benign production systems were retained in many traditional systems that for ecological, cultural or economic reasons were not effectively incorporated into the industrial model. Such systems sought to build on, rather than replace, natural ecosystems. Different modern approaches have focused on different aspects of ecological synergy, arising from differences in discipline, philosophy, problem focus or geographical conditions. Agroecology, permaculture (Mollison 1990), conservation agriculture (FAO 2001), agroforestry (Huxley 1999), organic agriculture (IFOAM 2000) and sustainable agriculture (Pretty 1999) have focused principally on maintaining the resource base for production, through managing nutrient cycles, protecting pollinators and beneficial micro organisms, maintaining healthy soils and conserving water. They sought to reduce the ecological 'footprint' of farmed areas and the damage to wild species from toxics, soil disturbance and water pollution, but most focused on farm scale action, rather than coordinating efforts among farmers and others to achieve demonstrable biodiversitybenefits at a landscape scale.

To protect wild fauna and flora, ecoagriculture landscapes must provide protection of nesting areas from disturbance, diverse perennial cover for protection from predators, adequate access to clean water throughout the year, territorial access between dispersed population groups to ensure provide minimum viable populations genetically and demographically, all season access to food from diverse sources, viable populations of predators and prey, healthy populations of other species with which they are interdependent(such as their pollinators), and biologically active soils. Many of these functions can be provided by healthy patches and networks of natural habitat, but production areas also play a critical role. To achieve these attributes in production areas, agricultural and conservation innovators are pursuing strategies such as minimizing agricultural pollution of natural habitats, managing conventional cropping systems in ways that enhance habitat quality, and designing farming systems to mimic the structure and function of natural ecosystems. A key challenge for farmers is to do so in ways that also maintain or increase agricultural output, reduce overall production costs or enhance the market value of their products in order to meet their broader livelihood needs while conserving biodiversity.

(i) Minimizing agricultural pollution of natural habitats

Reducing agrochemical use and livestock wastes in high input production systems can greatly benefit wildlife. For example, high nutrient or toxic run off into waterways (a problem for both natural and synthetic forms of nitrogen) can dramatically reduce aquatic biodiversity. Major advances have been made in methods to reduce and improve the efficiency of fertilizer use, through better timing and methods of application.Agricultural pesticides may also kill non target insects and weeds that constitute the foodbase for insect and grain eating species. Integrated pest management systems have effectively used varietal crop mixes, pest monitoring and management practices to reduce the need for pesticides (Kogan 1998). Cellular and molecular biology have been used to tailor pesticides to affect only specific pests. New ecological and biochemical research techniques are revealing an unexpected sophistication of host–pest relations thatcould revolutionize agricultural pest control in the future (Wittenberg and Cock, 2001). Pretty's (2005) meta review of farmer experience found gains in both productivity and biodiversity from reduced chemical use in developing counties. For example, the System of Rice Intensification mobilizes biological interactions in plant–soil systems, rather thanexternal inputs, to raise yields significantly while reducing costs (Uphoff et al., 2006).

Meanwhile, new whole farm planning approaches minimize run off of agrochemical and livestock waste into aquatic systems by improving storage systems, managing fields to improve infiltration and reduce run off, and establishing buffer zones to filter pollutants before they enter streams (Coombe 1996).

(ii) Managing production systems to enhance habitat quality

Farmers and conservationists have modified management of soil, water, fire and vegetation to transform crop fields into useful habitat for species, or to enhance their value as 'corridors' connecting natural habitat areas in the landscape (McNeely & Scherr 2003; Clay 2004). Buck et al. (2006) reviewed 79 studies where investigators quantified biodiversity (usually species richness) associated with 18 specific agricultural practices. The strategy most often correlated with the conservation of wild biodiversity was the maintenance of adjacent hedgerows, windbreaks or woodlots; 18 studies documented positive correlations with eight taxa. Organic agriculture was correlated with an increase in seven taxa in eight studies (Peach et al. 2001; Klein & Sutherland 2003; Kleijn et al. 2006). Shaded tropical crop production (especially coffee and cacao) had higher species richness of three higher taxa by eight different studies (Buck et al. 2007). Research has also found that many of these practices provide additional benefits to farmers, such as useful by products, reduced risk of crop loss during droughts, diversified food and income sources and reduced vulnerability to environmental risks. For example, following the October 1998 Hurricane Mitch (the worst natural disaster to strike Central America in 200 years), researchers found that farms using agro ecological practices suffered 58% less damage in Honduras, 70% less in Nicaragua and 99% less in Guatemala than those using conventional farming methods (Bunch 2000).

Impacts of conservation practices may be species specific. For example, cotton is an inhospitable habitat for many songbirds, particularly due to very high levels of pesticide use in conventional systems. But approaches such as conservation tillage and strip cover cropping reduce the ecological impact of cotton fields. Cederbaum et al. (2004) examined the effects of clover strip cover cropping with conservation tillage versus conventionally grown cotton with either conventional or conservation tillage on avian and arthropod species composition and field use in Georgia, USA. Strip cover fields had higher bird densities and biomass and higher relative abundance of arthropods than either conservation tillage or conventional fields. During migration and breeding periods, total bird densities on strip cover fields were 2–6 and 7–20 times greater than on conservation and conventional fields, respectively. Although the clover treatment attracted the highest avian and arthropod densities, conservation fields still provided more wildlife and agronomic benefits than conventional management. The reduction of inputs possible with the clover system allows farmers to reduce costs associated with conventional cotton production. Transgenic cotton has been developed to significantly reduce the need for pesticides, with observed benefits for biodiversity (Cattaneo et al., 2006).

The organic farming industry has only recently begun to develop standards that explicitly address conservation of wild biodiversity. But Hole et al. (2005) found that a wide range of taxa, including birds, mammals, invertebrates and arable flora can benefit from organic management through increases in abundance and/or species richness. Management practices, such as prohibition or reduced use of chemical pesticides and inorganic fertilizers, protection of non cropped habitats, and preservation of mixed farming, are particularly beneficial for farmland wildlife. Though yields from organic systems are still often lower than those in conventional systems, the gap is narrowing and research is accumulating that shows how agricultural production systems primarily or exclusively dependent on organic inputs can produce superior agronomic and economic results (Uphoff et al. 2006).

Carefully targeted management practices applied to relatively small areas of cropped or non cropped habitats within conventional agriculture may also provide valuable biodiversity benefits (Trewavas 2001). Weibull et al. (2003) found that wild species richness generally increased with landscape heterogeneity on a farm scale, and habitat type had a major effect on species richness for most groups, with most species found in pastures and leys (lands temporarily sown with grass). The level of motivation of the farmer to maintain biodiversity on the farmstead was more predictive of biodiversity outcomes than specific practices.

(iii) Modifying farming systems to mimic natural ecosystems

From a wild biodiversity conservation perspective, the ideal agricultural production systems for ecoagriculture landscapes mimic the structure and function of natural ecosystems (Leakey 1999; Lefroy et al. 1999; Jackson & Jackson 2002; Blann 2006). In humid and sub humid forest ecosystems, farms would mimic forests, with productive tree crops, shade loving understory crops and agroforestry mixtures; in grassland ecosystems, production systems would rely more on perennial grains and grasses, and economically useful shrubs and dry land tree species. Annual crops would be cultivated in such systems, but as intercrops, or monoculture plots interspersed in mosaics of perennial production and natural habitat areas. Domesticated crop and livestock species diversity would be encouraged at a landscape scale, and intra species genetic diversity would be conserved in situ at least at an ecosystem scale, to ensure system **resilience** and ecological diversity.

Multi story agro forest systems, tree fallows and complex home gardens are especially rich in wild biodiversity (Cairns & Garrity (1999); Schroth et al. 2004; Leakey & Tehoundjeu 2001). For example, Siebert (2002) found that canopy height, tree, epiphyte, liana and bird species diversity, vegetation structural complexity, per cent ground cover by leaf litter, and soil calcium, nitrate nitrogen and organic matter levels in Top soils were all significantly greater in shaded than in sun grown farms, while air and soil temperatures, weed diversity and per cent ground cover by weeds were significantly greater in sun farms.

Recent research in Central America has identified polyculture combinations and management systems that significantly improve the productivity of coffee, cocoa, banana, timber and other commercial tree products in these complex systems (e.g. Beer et al. 2000). New and improved perennial crops can substitute for products now provided by annuals, such as fruits, leafy vegetables, spices and vegetable oils. Perennial crops can be more resilient and involve less soil and ecosystem disturbance than annual crops, and provide much greater habitat value, especially if grown in mixtures and mosaics (Jackson & Jackson 1999; Leakey & Tchoundjeu 2001). Breeding efforts are also underway to perennialize annual grains and to mimic ecosystem functions of natural grasslands; in some cases, yields are becoming competitive with conventional varieties (de Haan et al. 2007). This is a significant research opportunity.

Increased demand for livestock products in turn raises demand for animal feed, including for higher quality pastures, fodder or inputs for concentrates. While historically low grain prices have meant that corn and soy have been dominant feedstock during the past few decades, alternatives abound, including perennial grass, shrub and tree species that can be grown more sustainably in marginal lands, as industrial processes adapt. Moreover, the future of industrial type intensive, grain fed livestock production is uncertain in the face of emerging zoonotic infectious diseases and associated pollution, opening more economic opportunity for substitutes from rotational grazing and even pastoral systems (Nierenberg 2005). Crops for bio fuels are poised to become one of the fastest growing segments of agricultural production, and although short term investments have favored annual crop sources in the developed world (as a way to absorb subsidy driven surpluses), grasses, shrubs and tree sources may be more economic and sustainable options once the technical challenges of processing cellulosic sources are overcome (Ruark et al. 2006).

(iv) Major gaps

(a) The development of agricultural practices and systems that explicitly support wild biodiversity is in its infancy.

Buck et al. (2006) highlight numerous critical knowledge gaps, especially knowledge about the link between diversity and ecosystem function, and the relationships between below and above ground biodiversity. Methods being used to assess biodiversity impacts are inadequate, and generally fail to evaluate the impact on regional or global diversity, or to interpret the significance of an individual member of a species found at a particular site. Researchers still find it difficult to link plot based analysis with landscape scale impacts (Tomich et al. 2004). Redford & Richter (1999) propose that researchers much more systematically assess the impact of different resource management options on specific components of biodiversity (the function, structure and composition of communities/ecosystems, populations/species, genetic diversity and option space; Redford & Richter 1999; Swift et al. 2004). Then, where eco agriculture systems are successful in increasing populations of wild species, new methods for managing them may be needed to minimize conflicts. The major gap is the miniscule level of international and national public investment in research documenting and evaluating existing eco agriculture production systems, or in pursuing agricultural and conservation research to improve biodiversity supporting and financially viable production systems.

(b) Managing eco agriculture landscapes for both production and conservation

The eco agriculture approach encompasses both biodiversity friendly agricultural production systems and practices, and their management in mosaics with natural areas and other landscape features to meet conservation, livelihood and production goals. One premise of eco agriculture is that ecosystem services can come from both production and conservation areas, especially if they are coordinated and managed for that purpose. Improved tools, greater demand for landscape scale action and reassessment of long sustained traditional agro ecosystems, have led to substantial progress over the past two decades in laying out the basic parameters for biophysical management of eco agriculture landscapes, if not location specific guidance.

(i) **New tools for landscape assessment:** Despite the importance of agricultural landscapes for biodiversity conservation, only a small fraction of published conservation biology studies has been undertaken in agricultural landscapes (Buck et al. 2004), so developing a baseline for assessing change is difficult. Most studies of the biodiversity impacts of particular agricultural practices and even the work of biodiversity oriented groups like the Rainforest Alliance have focused on farm level indicators. Meanwhile, a review of basic biodiversity research found very little empirical data on the contributions of wild species and natural ecosystem conditions to agricultural productivity (Jackson et al. 2005). Landscape ecology has provided us with the analytical language and tools to systematically examine the interactions between farmed and unfarmed areas (Forman 1995; Wojtkowski 2004).

(ii) **Maintaining natural habitats for terrestrial species in agricultural landscapes:** A common goal in eco agriculture landscapes is to conserve a broad range of terrestrial species native to the area. This includes species that are relatively **resilient** to habitat fragmentation and agricultural land use, as well as species that are rare or locally or globally threatened, and those that require larger extensions of minimally disturbed habitat. The prospects for achieving this in agricultural landscapes depends on the degree of fragmentation and functional connectivity of natural areas, the habitat quality of those areas, the habitat quality of the productive matrix and the behaviour of farmers.

For many commercial crop monocultures, leaving field margins uncultivated for habitat protection does not reduce total yields, as inputs were applied more economically on the rest (Clay 2004). However, landscape scale interventions specifically designed to protect habitats for biodiversity (that include but coordinate and go beyond farm and plot specific interventions) are much more effective. A recent review of evidence from North America on how much habitat is 'enough' in agricultural landscapes (Blann 2006) concluded that strategies need to consider habitat needs within the landscape history and context. Adequate habitat patch size and connectivity must be maintained, but 'adequate' must be considered in relation to matrix influence and patch condition (sinks and ecological traps, patch location and configuration, edge effects and boundary zones).Smaller patches of natural habitat may be sufficient if adjacent agricultural patches are ecologically managed. Based on studies from Central America, Harvey et al. (2005) conclude that landscape connectivity between large patches of forest can be effectively maintained through retention of tree cover on the farm, such as live fences, windbreaks, and hedges in grazing lands and agricultural fields. Sayer & Maginnis (2005) describe effective approaches for forest landscape restoration in mixed use mosaics.

(iii) **Optimizing agriculture–natural habitat interactions in landscape mosaics:** Biologically diverse agricultural systems and landscapes can contribute to control of pests and diseases, provide new economic species, and buffer environmental changes and challenges (Jackson et al. 2005; Thompson et al. 2007). Ricketts (2004) investigated the role of tropical forest remnants as sources of pollinators for surrounding coffee crops in Costa Rica, observing bee activity and pollen deposition rates at coffee flowers, including 10 species of native bees and the introduced honeybee, *Apis mellifera*. Bee species richness, overall visitation rate and pollen deposition rate were all significantly higher in sites within approximately 100 m of forest fragments than in sites farther away. The vast majority of pollination in coffee plantations more than 100 m from a forest was by the introduced honeybee. Forest fragments near coffee plantations increased both the amount and stability of pollination services by reducing dependence on a single introduced species. Kremen et al. (2002) found similar results for pollinators of watermelon fields near and far from natural woodlands in California.

(iv) **Major gaps:** The past two decades have revolutionized the potential for landscape scale assessment and scientific understanding of the ecological functioning of diverse types of agricultural landscapes. A framework for considering key management guidelines and broad parameters is now in place, but empirical or even ecological modeling evidence needed for managing eco agriculture landscapes (e.g. size and shape of natural areas required to sustain ecological functions, impacts on agricultural productivity of natural vegetation and species) is lacking. Agreed methods do not yet exist for integrated monitoring of livelihood, biodiversity and agricultural outcomes at a landscape scale although this challenge is being taken up (Scherr & Meredith 2005; Buck et al. 2007). Rigorous understanding of the potential benefits and costs for agriculture of associated wild flora and fauna, and key ecosystem services, is lacking (Jackson et al. 2005).

Molden et al. (2007) highlights numerous practices to manage irrigation water in ways that also support biodiversity. But these are not widely implemented because they are not part of the institutions, incentive structures and education related to irrigation. Little of the new science has been shared with farmers or even with agronomists and other specialized agricultural scientists and technicians. The science is often missing that informs real life innovations that local people can make to modify ecological impacts of management activities. Technical assistance services for farmers rarely address landscape management issues.

(c) Achieving financial viability of eco agriculture landscapes

Investing and engaging in eco agriculture systems will require that all key elements—farm production, nature conservation and associated institutions for collective landscape management—be adequately financed. If eco agriculture systems are to be widely adopted around the world, then incomes (defined to include not only cash but also other livelihood components) for farmers in those systems need to be at least as high, or higher, than in less biodiversity friendly production systems, and other non monetary benefits will be key. Market based innovations could provide many opportunities for scaling up eco agriculture or the possibilities of farm stewardship which we want emphasize in this thesis.

(i) **Making eco agriculture systems more profitable for farmers and investors:** Contrary to common assumptions, farmers and their communities often have strong economic and social rationales for supporting biodiversity conservation: to reduce production costs, raise or stabilize yields; improve product quality; protect their right to farm/ herd/harvest wild products in and around protected areas; comply cost effectively with environmental regulations; conserve biodiversity and ecosystem services critical to their own livelihoods; access product markets that require biodiversity friendly production systems; earn payments for ecosystem services; or conserve species and landscapes of special cultural, spiritual or aesthetic significance to them. Many eco agriculture systems are, in fact, more profitable or less risky than alternatives. McNeely & Scherr (2003) present 28 examples that clearly demonstrated positive economic benefits, and another five cases had a neutral impact on incomes (despite major benefits for wild biodiversity).

Pearce (2005) documents financial and economic contributions of ecosystem services and biodiversity to poverty reduction and vice versa. Communities are organizing themselves regionally to improve market linkages, reduce marketing costs and connect directly with buyers. Communities need to understand and meet the quality and time demands of interested buyers and to enter into and fully respect commercial contracts. They also need technical assistance to improve product quality and manage commercial contracts, and gain access to harvest finance and credit for post harvest product processing and handling facilities/technologies. Development of innovations at all points in the marketing chain can reduce costs for trading, storage, transport, bulking, grading, etc. and thus improve returns from marketing products from polycultures and multi product landscapes.

(ii) **Develop product markets that reward ecosystem stewardship:** New market niches are beginning to develop for agricultural products that are certified to be 'green'. Producers or products are certified by independent third parties to have positive or neutral effects on biodiversity, based on criteria such as reduced agrochemical pollution, protection of natural areas, use of production practices that do not interfere with key natural processes or species lifecycles and participation in the development of landscape scale wildlife corridors (Millard 2007). New markets are also developing for products based on sustainable harvest of wild species, or on the domestication of wild species (such as extracts, spices, medicinal plants, construction materials and fruits; STCP 2005). The use of marketing labels for agricultural products coming from particular geographical regions, originally focused on quality, culture and taste, is being adopted for products labeled as supporting conservation of high biodiversity value landscapes. Demand is growing as the food industry becomes more sensitive to reputation issues around environment, and advocates promote new institutional procurement policies (Rainforest Alliance 2006), although consumers remain motivated more by human health related issues. Concerns about bio terrorism and health, combined with low cost monitoring technologies, could enable farm to consumer farm product tracking to become more common in high and middle income countries, reducing the relative costs of managing value chains for eco certified products.

(v) **Reward farmers and farming communities for ecosystem services:** A major potential driver for eco agriculture landscapes is payments to farmers or herders/ranchers and their communities for conserving biodiversity important to outsiders, and for conserving other ecosystem services using management practices that also conserve biodiversity. Such compensation currently takes various forms, including payments for access to species or habitats (e.g. research permits; hunting, fishing or gathering permits for wild species; or ecotourism); payments for biodiversity conservation management (e.g. conservation easements, land leases, conservation concessions or management contracts); tradable rights under 'cap and trade' regulations (e.g. wetland mitigation credits, tradable development rights and biodiversity offset credits) and support for biodiversity conserving businesses (e.g. business investments or eco labeling of green products; Scherr et al. 2007a).

(vi) **Major gaps:** New mechanisms have arisen over the past decade to reward and finance biodiversity conservation and biodiversity friendly agriculture, but most of these are modest in scale or modestly effective at landscape scale. Little research has been done on the structures and institutions in product market chains that facilitate biologically diverse production. Nor has any systematic assessment been done of overall agricultural investment and finance and how it might be shaped to better support biodiversity. Certification processes for agricultural products (and wild products sustainably harvested) can be expanded, streamlined, designed for landscape scale impact, enable low income people to participate (IITA 2001; Molnar 2003). Still, most demand in developing countries is for domestic markets and seeking lowest cost supply, so it is crucial to focus on reducing costs across the market value chain (not only at the level of the producer).

ACHIEVING FARM STEWARDSHIP AT A GLOBALLY MEANINGFUL SCALE

This review found many examples of apparently successful approaches linking biodiversity conservation with sustainable agriculture. On the other hand, the current knowledge base and institutional arrangements are clearly inadequate to meet the objectives noted above across diverse ecosystems and production systems. To enable Farm Stewardship approaches to expand to a globally significant scale will require at least three elements: new knowledge, institutional capacity, and an enabling policy and market environment.

Produce and share knowledge for Farm Stewardship

The challenge of shaping agricultural landscapes to meet joint production, conservation and livelihood goals will require a dramatic scaling up and refocusing of research, in national research systems, different international agricultural research centre, centres of conservation science, national academies of science, and universities. Priorities are to understand the interaction and dynamics of conservation and production areas; to develop production systems (including improved varieties of more diverse domesticated species) that explicitly meet soil stewardship, water stewardship, pesticide stewardship, biodiversity stewardship objectives and mimic natural ecosystems; and to make more elements of farming systems ecologically sustainable, including industrial processing, packaging, transport, etc. Eco agriculture systems that appear to be successful need to be fully documented, both in terms of landscape scale outcomes and specific interventions. Mapping of spatial overlays between important agricultural areas (in terms of national product supply or local livelihoods) and important biodiversity will be essential.

Build capacity for Farm Stewardship

Knowledge innovation systems need to be reshaped to provide services to rural resource stewards, and to accelerate exchange of practical knowledge among them and across sectors. Rural communities must be acknowledged as key stewards of biodiversity conservation, and professional conservation organizations, public agencies and others need to reorient their activities to reflect this reality and provide services to community based organizations, as well as to other stakeholder groups. Conservation organizations need to fully embrace farming partners, develop agricultural expertise and aggressively advocate for sustainable agriculture investment in coordination with conservation strategies.

Promote markets and policies that support Farm Stewardship:

Technical and local organizational opportunities and initiatives for **farm stewardship** are unlikely to be successful unless major policy barriers are removed, and supportive policies developed. To advocate for this agenda, beneficiaries of ecosystem services provided by agricultural landscapes, new economic actors in the product value chain and advocates for reinvigorated rural development need to form new political coalitions. In North America, Europe, Japan, Australia and many developing countries, shifting of government funds from agricultural commodity subsidies to payments for ecosystem services (including carbon emission offsets) in eco agriculture landscapes could provide initial funding to build institutions and farmer capacity for Farm Stewardship. It offers cost effective approaches for national investment strategies to achieve the Millennium Development Goals. Strategic changes in the food industry, institutional procurement, eco certification of agricultural products and financial investors' oversight of agricultural investments can be mobilized to shift financial incentives towards farm stewardship. At the international policy level, opportunities exist to integrate stewardship strategies into the work programmes of the international environmental conventions, and to ensure that rules of the World Trade Organization support eco agriculture landscapes.

Managing Food Production System for Stewardship

Managing food production system on a sustainable basis is one of the most critical challenges for the future of humanity, for the obvious reason that people cannot survive without food. Ecosystem health is both a "means" and an "end" to resilient crop and animal production. Being fundamentally dependent on the world's atmosphere, soils, freshwater, and genetic resources, these system generate some of the most ecosystem services on the planet. They are also the largest global consumers of land and water, the greatest threat to biodiversity through habitat change and invasive species, significant sources of air and water pollution in many locations, and major determinants of biogeochemical change from local to global scales (Matson et al. 1997, Vitousek et al.1997, Naylor 2000, Smil 2000). The inherent interplay between human welfare, food production, and the state of the world's natural resources underscores the need to manage these systems for resilience to anticipate change and shape it in ways that lead to the long run health of human populations, ecosystems, and environment quality.

The challenge of twenty first century is to develop food production systems in a way that will support rural incomes, enhance yield growth, utilize inputs efficiently (particularly water and added nutrients), minimize environmental impacts, and provide healthy diet for the human population (Conway 1997, Power 1999, Tilman et al. 2002, Robertson and Swinton 2005). There is no question that the productivity of food production systems must continue to increase in order to achieve any definition of sustainability, but it is not clear that the goal should be to maintain or augment productivity on existing system in all cases. Instead the goal should be to redesign systems to promote genetic and crop diversity, stability in the face of future shocks (e.g., climate change and associated pest predator impacts), and food security as defined by access to affordable food for all people at all times. In order to achieve this goal, effort should be directed toward both small scale farming systems that primarily meet local and regional demands, and large scale surplus system that meet national and global demands (Robertson and Swinton 2005).

The Perspective of Farm Stewardship

The diversity of agricultural system seen throughout the world and should not be considered as the typical agriculture of any particular region. The process of maintaining and managing the resilience and sustainability of one location do not necessarily apply to other areas with similar biophysical systems and also not transferrable across scales of production. The global food production system is complex.

Despite such complexity, the transition toward resilient food production systems will require three main steps (Lambin 2007). The first step involves a widespread recognition that the biophysical and socio economic conditions of the twenty –first century are markedly different than the past and that many crop and animal production system will need to be redesigned fundamentally. Some principle could help shape this process of redesign. For example, given the growing resource constraints on agriculture, it is time to focus more on adapting crops and animals to the local resource base and nutritional needs, rather than transforming the resource base to meet specific commodity production targets. Does it make sense to grow water intensive crops in irrigated – ecosystems or erosion – prone crops on hillsides? Or does it make sense to decouple crops from livestock and use large amounts of fertilizers to produce feed? The shift toward production systems that are more compatible with the human and natural resource base leads to another principle for redesign: the promotion of crop diversity on local to global scale. Creating a resilient path for food production systems worldwide begins, in the words of Wendell Berry (2005) "in the recognition and acceptance of limits." It also begins with a broad understanding of the economic, social, and climate forces that are changing the agricultural landscape.

In order to promote a vision of redesign operationally, the second step is that government policies influencing the trajectory of food production systems must be altered in order to re couple agriculture with its environmental support systems

(Robertson and Swinton 2005). For example, distorting policies that encourage crop or animal production that is incompatible with resource constraints should be discarded, and incentives to improve the efficiency of input use and eliminate external impacts from agriculture should be introduced.

The Third and final step toward the resilience transition is to promote the scientific and information tool kit that enables farmers to anticipate and respond to change within the local cultural context. The scientific tool kit is broad and includes emerging knowledge from a wide range of disciplines (e.g., genetics, agronomy, biology, hydrology, climate science, economics) to improve management practices, breeding efforts, cultivar development and deployment, local adaptation to climate change, water availability and options for income and nutritional enhancement. Advanced genetics will play a key role in identifying desired crop traits and increasing productivity of major and minor crops under evolving stresses; this field extend beyond the use of genetically modified organisms to include marker assisted breeding and bio informatics (Naylor et al. 2004). Integrated management practices that improve input use efficiency (water and nutrients) will be equally, if not more important for enhancing crop productivity, incomes and environmental services in all areas of the world (Cassman et al. 2003). Finally, the use of new information technology, such as remote sensing images and geographic information systems (GIS) to identify yield gaps, input use efficiencies, soil and water constraints, and climate projection on regional scales, could help transform the ability of farmer in both rich and poor countries over time (Cassman 1999, Lobell and Oritz Monasterio 2008).

The main point is that efforts are needed to develop the full toolkit as an integrated package throughout the world, thus helping to preserve options for future adaptation to change (Solow 1991). Such efforts will require the institutional commitment of international agriculture and development agencies and donors, as well as the commitment of the national governments, to reach farmers in all regions (Falcon and Naylor 2005). Many farmers are already redesigning their agricultural systems to meet local needs and constraints. It is now time for the global community to embrace a vision of redesign of food production systems to meet human needs without compromising future options, cultural integrity, and environmental quality.

The world food crisis that hit agricultural commodity markets in 2008 is a stark reminder that all countries need to build resilient food production systems.

Expected Outcome of the study

1. The present status of conservation practices followed and perceived stewardship possibilities among the farmer.
2. The indicator to measure the paradigm shift from exploitative agriculture to conservational agriculture in terms of selected variables.
3. The clandestine shift of practicing conservation to possible stewardship in terms of technology, resources and ecology.
4. An extension Policy generation , as applicable for the research locale, as well as similar ecological set up elsewhere for integrating conservation with stewardship through an approach of ecological resilience.

Research Question

1. Are the farmers really concerned about conservation and stewardship, and if so what are the cognitive, perceptual and operational levels?
2. What are the set of factors influencing and characterizing the nature of conventional conservational practices followed and serendipitous possibilities of stewardship?
3. Is there any paradigm shift from exploitative agriculture to conservative agriculture leaving a space for stewardship possibilities to dovetail productive farming with ecological resilience?
4. What are the extension and policy implication of the study?

Axiomatic statement

1. Individual vis a vis respondents have got unique or common socio ecological surroundings and they are supposed to respond from their unique or common socio ecological bondage and interaction. A baseline is the present standards or level from which they derived perception and respond to stimuli.
2. Every individual respondent in a functional social ecology is being impacted or responding to a set of characters, called variables. The variables for this study are objective and verifiable and follow a normal distribution. For conventional conservation approaches and techniques, respondents are constantly being characterized and keep characterizing the system in terms of objectively verifiable indicators for variables. The variables are mutually dovetailed in a causal or interactive relationship.

3. Each set of variables, when accepted for or made operative in a functional socio ecological system, are presenting both between and within interactions to ultimately characterize the system behavior. Prediction, estimation and elucidation are enacted based on the intensity, nature and direction of the interaction commencing both at inter and intra level.
4. The dents and direction of conservation are naturally attuned towards achieving a level of stewardship. This has been a shift from digitized conservation to operational stewardship by adding values and resilience to the classical conservational intervention and approaches. Natural resources cannot be conserved as such unless anthropogenic surroundings are organically orchestrated with extrapolation of ecological deliverables as well as services towards fulfilling human as well system needs.
5. Every empirical research, here it's on conservation and stewardship, is expected to generate adaptable, actionable and amenable policy as applicable to the micro level research locale for achieving resilient and sustainable ecosystem functioning.

4 CHAPTER

SOCIAL ECOLOGY AND SETTING

Research setting refers to the detail information of an area where the study has been conducted. The study area, generally comprises of a particular geographical area viz. a state, district, block, or gram panchayat area selected according to the convenience of the investigator who possess adequate knowledge regarding the location, communication facility etc. of the locality so that he/she can easily approach each and every corner of the area for data collection. Besides the investigators must also have the basic knowledge about the socio demographic background of the people so as to have an easy understanding of their knowledge, attitude and behavior. Thus, a study on Monsoon impact analysis certainly requires the information about the impact of monsoon on the economic, social and cultural aspects of people.

The present study was taken up at Panchkahaniya, Bhabanipur, and Satyapul , these three villages of Fatehpur gram panchayat in Haringhata block in West Bengal. A brief description of the state and district in general and the block and village in particular is given below.

Area of Study

The area of investigation of this study is situated in the state of West Bengal located in the eastern part of India. The state of West Bengal in eastern India has unique social and ecological backgrounds which influence the living standard and behavioral patterns of the people in many ways. The area of investigation belongs to the Haringhata block in Nadia district. The area of the study is Panchkahaniya, Bhabanipur, and Satyapul , these three villages of Fatehpur gram panchayat.

Profile of the State West Bengal

West Bengal was created as a one of the constituent state of the Indian union among the 29 states in India on 15th August, 1947 as the result of partition of the undivided

British Indian province of Bengal into West Bengal. West Bengal is situated in the north eastern part of India and lies between 21 37′ 27 10′ north latitude and 85 51′ 89 53′ east longitude.

It is flanked by the Bay of Bengal on the South, Sikkim on the north, Assam on the east and Bihar, Jharkhand and Orissa on the west. It also has common borders with the three neighboring countries viz. Nepal and Bhutan in the north and Bangladesh in the east. The tropic of cancer runs across the middle of the state and passes through four districts viz. Bankura, Burdwan, Nadia and Purulia. The state has a total geographical area of 88,752 sq. km. representing only 2.7% of the total area of India. It is territorially divided into 19 districts i.e. Coochbehar, Darjeeling, Jalpaiguri, Uttar Dinajpur,

Dakshin Dinajpur, Malda, Murshidabad Hoogly Howrah, Nadia, Birbhum Burdwm Bankura, Purulia, 24 Par2ana(N) 24 Pargana(S) Purba Midnapur, Pashim Midnapur and Kolkata. There are more than 341 Community Development bocks in West Bengal. There are 3354 Gram Panchayats and more than 42000 villages in West Bengal.

In 1947, when India became independent Bengal was partitioned between India India's share came to be known as West Bengal and Pakistan's share was called East Pakistan Later, the state of Cooch Behar, French enclave of Chandranagore and some parts of Bihar were added to West Bengal West Bengal represents the lands that possess a district culture with its indigenous art and crafts and make it an important part of the Indian Union.

Table 4.1. General Information of West Bengal

Total population	91,347,736
Male	46,927,3 89
Female	44,420,347
Population Growth	13.93%
Sex ratio	947
Population density	1029/ km
Literacy rate	77.08%
Male	82.67%
Female	71.16%
No of Sub divisions	66
No of blocks	341
No of Gram Panchayats	3354
Inhabited Villages	37945
Area (Sq. Km)	88752
% of Electrified Villages	83.6

Source: Office of the Register General and Census Commissioner 2011 Ministry of Home Affairs, GOI.

Climate of the state

West Bengal's climate varies from tropical savanna in the southern portions to humid subtropical in the north. The main seasons are summer, the rainy season, a short autumn, and winter. While the summer in the delta region is noted for excessive humidity, the western highlands experience a dry summer like northern India, with the highest daytime temperature ranging from 38 C (100 F) to 45 C (113 F). At night, a cool southerly breeze carries moisture from the Bay of Bengal. In early

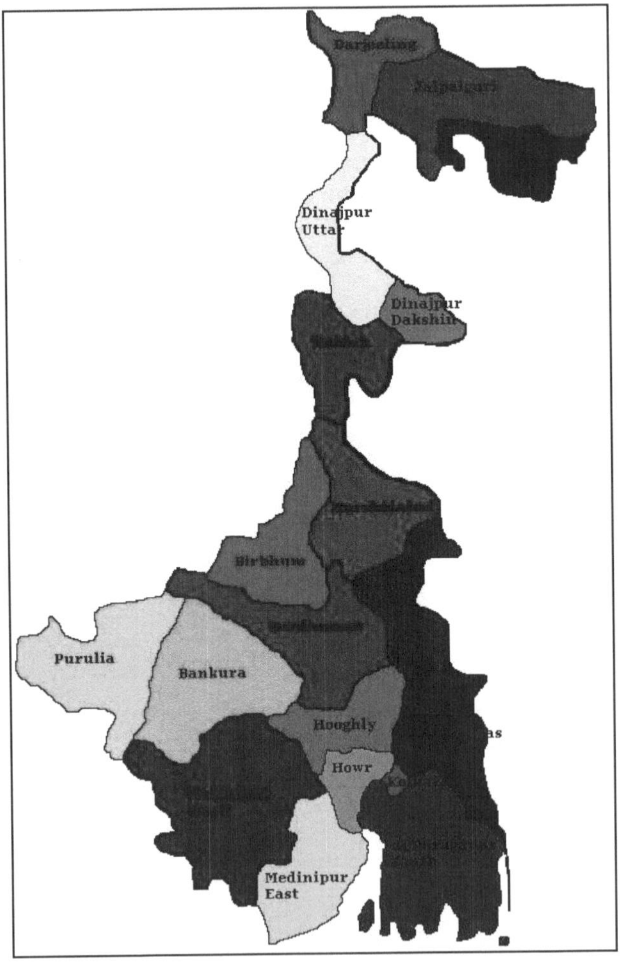

Fig 4.1. Map of West Bengal

summer, brief squalls and thunderstorms known as Kalbaisakhi, or Nor'westers, often occur. West Bengal receives the Bay of Bengal branch of the Indian Ocean monsoon that moves in a southeast to northwest direction. Monsoons bring rain to the whole state from June to September. Heavy rainfall of above 250 centimeters (98 in) is observed inthe Darjeeling, Jalpaiguri, and Cooch Behar district. During the arrival of the monsoons, low pressure in the Bay of Bengal region often leads to the formation of storms in the coastal areas. Winter (December–January) is mild over the plains with average minimum temperatures of 15 C (59 F). A cold and dry northern wind blows in the winter, substantially lowering the humidity level. The Darjeeling Himalayan Hill region experiences a harsh winter and occasional snowfall.(https://en.wikipedia.org/wiki/West_Bengal)

Profile of Nadia District

The district of Nadia is situated in the heart of the Bengal delta held within the arms of the Ganga, namely, the Bhagirathi on the West and Mathabhanga on the North. The entire district lies in the alluvial plain of the Ganga and its tributaries. Most districts in West Bengal take their name from the headquarters station of the district, but Nadia district takes its name not from Krishnagar, the headquarter but from Nadia or Nabadwip hallowed by the memory of Lord Shri Chaitanya Mahaprabhu who was born here on 18th February, 1486. The British district of Nadia was formed in 1787. The present district of Nadia after partition was formed by Notification No.545 GA dated 23rd February, 1948.

The district of Nadia is bounded on the north and north west by the district of Murshidabad. On the south east and east it is bounded by the Republic of Bangladesh. Inthe south and south east the district is bounded by the district of North 24 Parganas. The shape of the district is irregular, lying North to South. The district is about 46ft. above the mean sea level and the tropic of cancer divides the district into two parts.

For administrative convenience Nadia district is divided into four Subdivisions – Krishnagar Sadar, Ranaghat. Kalyani and Tehatta. The district has an area of 3927 sq kms having a population of 46, 04,827 as per Census 2001. Out of that SC & ST population are 13,65,985 and 1,13891 respectively. The district has 17 Panchayet Samities consist of 187 Gram Panchyets and 8 Municipalities. Total number of Police Stations in the district is 19. The density of population in this district is 1173 persons per sq km. Nadia district has 950 females as against 1000 males. The majority of the people of the district speak Bengali followed by Hindi, Santali and other. Religion wise about 73.75% are Hindus and 25.42% are Muslims. In the district of Nadia the percentage of literacy by sex is 72.30 (Male) & 59.60 (Female) as per Census 2001. The important rivers of the district are Bhagirathi, Churni, Mathabhanga, Ichamati and jalangi.

Krishnanagar is the most important town of this district, as it is the district headquarters of Nadia. It is on the bank of river Jalangi. The palace of Nadiaraj (Rajbari)built here is a prominent place of tourist attraction though the remnants of the past glory have been eroded and only a dilapidated structure of the exquisite places with carving on its inner walls exists today. Krishnagar is famous for world class clay models & terracotta works. It can claim a proud tradition for the sweets Sarpuria & Sarbhaja.

Table 4.2. General Information of Nadia District

Total population	5,167,600
Male	2,653,768
Female	2,513,832
Population Growth	12.22%
Sex ratio(per 1000)	947
Population density	1300/ km^2
Literates Rate	74.97%
Male	78.75%
Female	70.98%
No of Sub divisions	5
No of Development blocks	17
No of Gram Panchayats	187
Inhabited Villages	2639
Area (Sq. Km)	3,927

Source: Office of the Register General and Census Commissioner 2011 Ministry of Home Affairs, GOI.

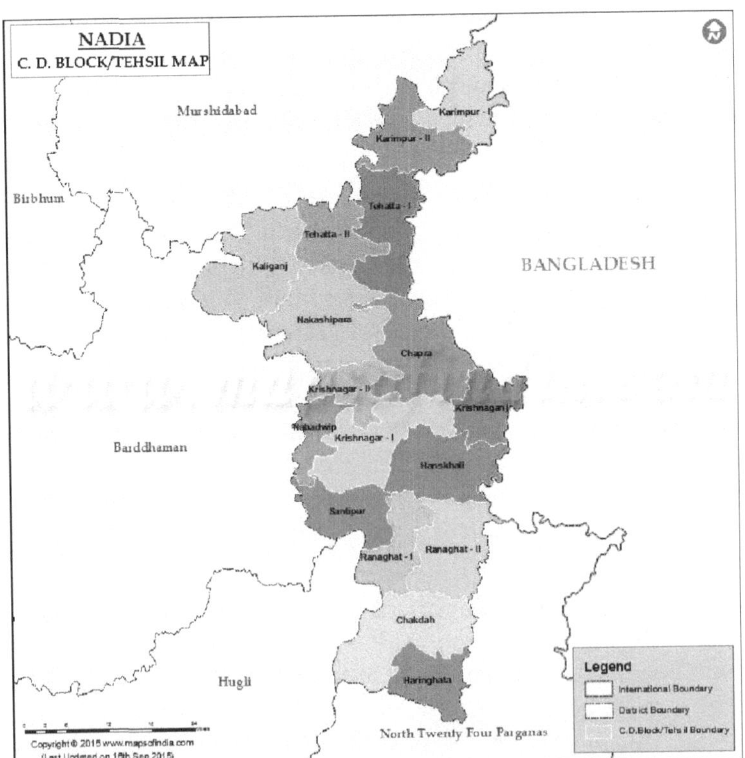

Fig 4.4. Map of Nadia District

Profile of Haringhata Block

Haringhata is a town in Kalyani subdivision of Nadia district in the state of west Bengal of India.

Location

Haringhata is located at 22 95′N 88 57′E.

It has an average elevation of 10 meters. Haringhata community development block has an area of 168.59 Km^2.

Language

Bengali is the local language in these areas.

Table 4.3 General Information about Haringhata block

Block area	170.32 sq.Km
No. of gram panchayets	10
No. of gram sabhas	129
No. of Mouza No. of police stations	87 1
Elevation	13 m (43 ft)
Total Population (2011)	231,068
Density	1,400/km^2 (3,500/sq mi)
Languages	Bengali, English
Total literates	171,144 (82.15%)

Geography

Haringhata Block is bounded by Chakdaha Block in the north and west, Bangaon and Gaighata Blocks, in North 24 Parganas district, in the east, Amdanga, Habra I and Habra II Blocks, in North 24 Parganas district, in the south. Nadia district is mostly alluvial plains lying to the east of Hooghly River, locally known as Bhagirathi. The alluvial plains are cut across by such distributaries as Jalangi, Churni and Ichhamati. With these rivers getting silted up, floods are a recurring feature. Haringhata Block has an area of 170.32 km^2. It has 1 panchayat samity, 10 gram panchayats, 176 gram sansads (village councils), 87 mouzas and 82 inhabited villages. Haringhata police station serves this block. Headquarters of this Block is at Subarnapur. It is located 56 km from Krishnanagar, the district headquarters.

Gram panchayats of Haringhata block/ panchayat samiti are: Birohi 1, Birohi 2, Fatepur, Haringhata 1, Harinhgata 2, Kastodanga 1, Kastodanga 2, Mollabelia, Nagarukhra 1 and Nagarukhra 2.

Demographics

(i) Population

As per the 2011 Census of India, Haringhata Block had a total population of 231,068, of which 207,459 were rural and 23,609 were urban. There were 118,709 (51% males and 112,359 (49%) females. Scheduled Castes numbered 73,755 (31.92%) and Scheduled Tribes numbered 11,218 (4.85%).

As per the 2001 census, Haringhata block had a total population 207,867, out of which 108,011 were males and 99,856 were females. Haringhata block registered a population growth of 16.92 per cent during the 1991 2001 decade. Decadal growth for the district was 19.51 per cent. Decadal growth in West Bengal was 17.84 per cent.

There are four census towns in Haringhata Block (2011 census figures in brackets) : Haringhata (3,989), Simhat (6,949), Subarnapur (5,759) and Digha (6,916).

Large villages (with 4,000+ populations) in Haringhata Block were (2011 census figures in brackets): Chandirampur (4,740), Narayanpur (4,548), Uttar Datta Para (5,430), Uttar Brahmapur (5,371), Subuddipur (6,051), Sekendarpur (7,917), Kathdanga (7,417), Mollabalia (5,889), Ganguria (4,291), Fatepur (5,093), Barasat (4,046), Ukrah (13,548), Goaldob (4,883) and Dighalgram (5,390).

Other villages in Haringhata Block include (2011 census figures in brackets): Birohi (3,721).

(ii) Literacy

As per the 2011 census, the total number of literates in Haringhata CD Block was 171,144 (82.15% of the population over 6 years) out of which males numbered 93,313 (87.11% of the male population over 6 years) and females numbered 77,831 (76.90% of the female population over 6 years). The gender disparity (the difference between female and male literacy rates) was 10.21%.

(iii) Language and religion

Bengali is the local language in these areas.

In the 2011 census, Hindus numbered 164,263 and formed 71.09% of the population in Haringhata Block. Muslims numbered 65,148 and formed 28.19% of the population. Christians numbered 460 and formed 0.20% of the population. Others numbered 1,197 and formed 0.52% of the population.

In the 2001 census, Hindus numbered 151,379 and formed 72.79% of the population of Haringhata CD Block. Muslims numbered 55,885 and formed 26.97% of the population. In the 1991 census, Hindus numbered 130,125 and formed 73.19% of the population of Haringhata CD Block. Muslims numbered 47,284 and formed 26.60% of the population.

Rural Poverty

The District Human Development Report for Nadia has provided a CD Block wise data table for Modified Human Vulnerability Index of the district. Haringhata

CD Block registered 31.59 on the MHPI scale. The CD Block wise mean MHVI was estimated at 33.92. A total of 8 out of the 17 CD Blocks in Nadia district were found to be severely deprived when measured against the CD Block mean MHVI Karimpur I and Karimpur II (under Tehatta subdivision), Kaliganj, Nakashipara, Chapra, Krishnanagar I and Nabadwip (under Krishnanagar Sadar subdivision) and Santipur (under Ranaghat subdivision) appear to be backward.

Economy

(i) Livelihood

In Haringhata CD Block in 2011, amongst the class of total workers, cultivators formed 18.42%, agricultural labourers 35.60, household industry workers 3.79% and other workers 42.19%.

The southern part of Nadia district starting from Krishnanagar I down to Chakdaha and Haringhata has some urban pockets specializing in either manufacturing or service related economic activity and has reflected a comparatively higher concentration of population but the urban population has generally stagnated. Nadia district still has a large chunk of people living in the rural areas.

(ii) Infrastructure

There are 82 inhabited villages in Haringhata Block. 100% villages have power supply and drinking water supply. 13 Villages (15.85%) have post offices. All 82 villages (100%) have telephones (including landlines, public call offices and mobile phones). 64 villages (78.05%) have a pucca approach road and 30 villages (38.59%) have transport communication (includes bus service, rail facility and navigable waterways). 6 villages (7.32%) have agricultural credit societies and 8 villages (9.76%) have banks. It should, however, be noted that although 100% villages in Nadia district had power supply in 2011, a survey in 2007 08 revealed that less than 50% of households had electricity connection. In rural areas of the country, the tube well was for many years considered to be the provider of safe drinking water, but with arsenic contamination of ground water claiming public attention it is no longer so. Piped water supply is still a distant dream. In 2007 08, the availability of piped drinking water in Nadia district was as low as 8.6%, well below the state average of around 20%.

(iii) Agriculture

Although the Bargadari Act of 1950 recognised the rights of bargadars to a higher share of crops from the land that they tilled, it was not implemented fully. Large tracts, beyond the prescribed limit of land ceiling, remained with the rich landlords. From 1977 onwards major land reforms took place in West Bengal. Land in excess of land ceiling was acquired and distributed amongst the peasants. Following land reforms land ownership pattern has undergone transformation. In 2013 14, persons engaged in agriculture in Haringhata Block could be classified as follows: bargadars 2.44%, patta (document) holders 7.95%, small farmers (possessing land between 1 and 2 hectares) 6.18%, marginal farmers (possessing land up to 1 hectare) 33.53% and agricultural labourers 49.09%.As the proportion of agricultural labourers is very high, the real wage in the agricultural sector has been a matter of concern.

Haringhata CD Block had 207 fertiliser depots, 40 seed store and 44 fair price shops in 2013-14.

In 2013-14, Haringhata CD Block produced 72,866 tonnes of Aman paddy, the main winter crop from 24,942 hectares, 20,348 tonnes of Boro paddy (spring crop) from 5,720 hectares, 5,413 tonnes of Aus paddy (summer crop) from 2,013 hectares, 4,317 tonnes of wheat from 1,315 hectares, 42,799 tonnes of jute from 2,383 hectares and 16,616 tonnes of potatoes from 536 hectares. It also produced pulses and oilseeds.

In 2013-14, the total area irrigated in Haringhata Block was 967 hectares, out of which 89 hectares were irrigated by river lift irrigation, 830 hectares by deep tube wells and 48 hectares by shallow tube wells.

Haringhata Dairy Farm

In order to supply milk to the urban population in West Bengal, the first dairy plant was installed in 1950 at Haringhata by the Directorate of Dairy Development, under the aegis of Animal Resources Development Department, Government of West Bengal. It started with a production of 1.5 thousand litres of bottled milk per day in 1950 51. With a network of two dairy plants, seven milk colony units and 19 Milk Collection cum Chilling stations named as "Greater Calcutta Milk Supply Scheme" the production had gradually gone up to 210 thousand litres per day by 1980s. (Directorate of Dairy Development, under the aegis of Animal Resources Development Department, Govt. of West Bengal)

Banking

In 2013-14, Haringhata Block had offices of 12 commercial banks and 2 gramin banks.

Transport

Haringhata CD Block has 4 originating/ terminating bus services. The nearest railway station Kanchrapara is 16 km and Kalyani Railway Station is 19 km from Block headquarters.

NH 12 (old number NH-34) passes through this block. SH 1, running from Bangaon (in North 24 Parganas district) to Kulpi (in South 24 Parganas district) passes through this Block.

Education

In 2013 14, Haringhata Block had 120 primary schools with 9,251 students, 11 middle schools with 866 students, 3 high schools with 1,819 students and 18 higher secondary schools with 23,167 students. Haringhata Block had 1 general college with

3,233 students, 2 technical/ professional institutions with 1,880 students and 453 institutions for special and no normal education with 10,874 students. In Haringhata Block, amongst the 82 inhabited villages, 4 villages did not have any school, 21 had more than 1 primary school, 16 had at least 1 primary and one middle school and 14 had 1 middle school and 1 secondary school.

Healthcare

In 2014, Haringhata Block had 1 rural hospital, 4 primary health centres and 2 private nursing homes with total 96 beds and 9 doctors (excluding private bodies). It had 30 family welfare sub centres. 4,907 patients were treated indoor and 357,557 patients were treated outdoor in the hospitals, health centres and sub centres of the Block. Haringhata Rural Hospital at Mohanpur has 25 beds. There are primary health centres at Birohi (6 beds), Kastadanga (6 beds) and Nagarukhra (10 beds). Haringhata Block is one of the areas of Nadia district where ground water is affected by high level of arsenic contamination. The WHO guideline for arsenic in drinking water is 10 mg/ litre, and the Indian Standard value is 50 mg/ litre. All the 17 blocks of Nadia district have arsenic contamination above this level. The maximum concentration in Haringhata Block is 865 mg/litre.

Profile of the Village About Panchkahania

According to Census 2011 information the location code or village code of Panchkahania village is 322379. Panchkahania village is located in Haringhata Tehsil of Nadia district in West Bengal, India. It is situated 5.6km away from sub district headquarter Subarnapur and 59km away from district headquarter Krishnanagar. As per 2009 stats, Fatepur is the gram panchayat of Panchkahania village.The total geographical area of village is 138.11 hectares. Panchkahania has a total population of 1,400 peoples. There are about 335 houses in Panchkahania village. Kalyani is nearest town to Panchkahania which is approximately 10km away. Total Population of the vilaage is 1400, where male population is about 751 and female population is about 649. Public bus service is available within 5 km distance and railway station within 10 km distance.

About Satyapul

According to Census 2011 information the location code or village code of Satyapul village is 322382. Satyapul village is located in Haringhata Tehsil of Nadia district in West Bengal, India. It is situated 4.4km away from sub district headquarter Subarnapur and 61.2km away from district headquarter Krishnanagar. As per 2009 stats, Fatehpur is the gram panchayat of Satyapul village. The total geographical area of village is 165.04 hectares. Satyapul has a total population of 1,575 peoples. There are about 392 houses in Satyapul village. Habra is nearest town to Satyapul which is approximately 10km away. Total Population of the village is 1575, where male population is about 792 and female population is about 783. Public bus service is available within 5 km distance and railway station within 5 km distance.

About Bhabanipur

According to Census 2011 information the location code or village code of Bhabanipur village is 322381. Bhabanipur village is located in Haringhata Tehsil of Nadia district in West Bengal, India. It is situated 3.6km away from sub district headquarter Subarnapur and 53km away from district headquarter Krishnanagar. As per 2009 stats, Fatepur is the gram panchayat of Bhabanipur village. The total geographical area of village is 116.49 hectares. Bhabanipur has a total population of 793 peoples. There are about 196 houses in Bhabanipur village. Habra is nearest town to Bhabanipur which is approximately 103km away. Total Population of the village is 793, where male population is about 408 and female population is about 385. Public bus service is available within 5 10 km distance and railway station within 10 km distance.

5

CHAPTER

RESEARCH METHODOLOGY

The deliberation on the methodology has been made to understand the concept, methods and techniques which are utilized to design the study, collection of information, analysis of data and interpretation of the findings for revelation of truths and formulation of theories. These chapter deals with the method and a procedure used in the study an consist of eight main parts

A. Locale of research.

B. Pilot study.

C. Sampling Design.

D. Item analysis and experts rating for devising scoring technique on dependent variables.

E. Empirical measurement of the variables.

F. Preparation of interview Schedule.

G. Pre testing of Interview Schedule.

H. Techniques of Data collection.

I. Statistical Tools used for Analysis of Data.

LOCALE OF RESEARCH

Panchkahaniya, Bhabanipur, Satyapul these three villages of the Haringhata block of Nadia district in West Bengal was randomly selected for the study. The area has been selected for the study because of – a) There is sample scope for collecting relevant data for the present study, b)Acquaintance with the local people as well as local language, c)The concern area was very easily accessible to the researcher in terms of place of residence, d)the area was very easily accessible to the researcher in terms of transportation and e) The closer familiarities of the student researchers with the area, people, officials and local dialects.

PILOT STUDY

Before taking up actual field work a pilot study was conducted to understand the area, its people, institution, communication and extension system and the knowledge, perception and attitude of the people towards climate changes concept.

SAMPLING DESIGN

Purposive as well as simple random sampling techniques were adopted for the study.

Table 5.1 Sampling Techniques and Sampling Design

STEPS	LEVEL	CRITERIA	APPROACH
Step 1	State(1) West Bengal	Need for soil water biodiversity conservation is well discernible. Huge population pressure and urbanization generating share on land resources. The state is undergoing faster rate of farm mechanization and chemicalization.	Purposive
Step 2	District(1) Nadia	Access. Predomination of erodibility of farm resources. Farmers are mostly small and marginal, cropping intensity is one of the highest in the state.	Purposive
Step 3	Block(1) Haringhata	Representative to the problem statement. Good accessibility. Prevalence of differential modernization. Multilayer character. Domain variability.	Purposive
Step 4	Village(3) Panchkahaniya, Bhabanipur, Satyapul	Predomination of the socio ecological parameter selected for the study. Upcoming stewardship functions , e.g., mulching, land shaping, biogas, withdrawal for water intensive Boro rice cultivation by welcoming leguminous and oilseed crops. Institutional Relation with the selected village.	Randomly
From each village 50 respondents were selected randomly which are well representative of these village farm families. So total number of respondents is 50*3=150.			

ITEM ANALYSIS AND EXPERTS RATING FOR DEVISING SCORING TECHNIQUE ON DEPENDENT VARIABLES

Since the topic of the research has been so new and unique, the present researcher and as per guidance of the supervisor has generated a score of marker questions to be rated on four point scales by the selected experts comprising of eminent scientist in the concerned areas and extension practitioner widely exposed to the domain. The steps wereas follows

a) Random selection of item.

b) Screening and scrutinizing of the items based on applicability and relevancy.

c) Finalization of marker item / statements to be rated on four point scale.

d) Selection of the experts and seeking their consents.

e) Sending the item to the selected experts and collecting their feedback (80 percent of the feedback from the experts were available).

g) Respective item values were summated.

g) 50 percent of the items were retained and rests were discarded.

h) Through before after test on 20 selected respondents, the r value was calculated and it was found significant.

i) The final marker statement have thus been finally selected and applied for data collection.

Details of the items used on six stewardship and conservation are given in theAnnexure II.

EMPIRICAL MEASUREMENT OF THE VARIABLES

After reviewing various literatures related to the field of study and consultation with respected chairman Advisory Committee and other experts, a list of variables was prepared.

Table 5.2. List of Dependent variables

Sl. No.	Dependent Variables	Notations
1.	Application of organic manure	Y1
2.	Rice stubble height left in the Field	Y2
3.	Crop residue left in the field	Y3
4.	Soil conservation level	Y4
5.	Crop Biodiversity conservation	Y5
6.	Climate resilient technology	Y6
7.	Ecosystem output	Y7

Application Of Organic Manure (Y1)

Organic manure has been perceived as organic source of plant nutrient called manure either produced at domestic level or produced from elsewhere. It is measured in terms of kg/ha applied in the field.

Rice Stubble Height Left In The Field (Y2)

After harvesting some part of the paddy straw, the vegetative plant parts are leftin the field. These are called stubbles and important source organic manures. The higher is the length of stubble, left, the more is the volume of the organic manure.

Here it has been measured in terms of inches by using scale and taking the average length of five stubbles collected from different parts of the field.

Crop residue left in the field (Y3)

Crop residue means the unused or uneconomic parts of the plants (Vegetables, pulsed etc.). These are the by product and also used as important source of organic manure. These are measured in terms of kg/ha.

Soil conservation level (Y4)

Soil conservation level is an input indicator for estimating stewardship level. Soil may be conserved by traditional community practices or by institutional practices in the form of recommendation. Here it has been measured as Percentage of area under Mulching, land shaping, contour bunds, paira cropping divided by area under cultivation.

Crop Biodiversity Conservation (Y5)

Crop biodiversity indicates the distribution and intensity of plants/ crops / vegetables etc. per unit area. Here Simpson index was used to measure it.

A community dominated by one or two species is considered to be less diverse than one in which several different species have a similar abundance.

Simpson's Diversity Index is a measure of diversity which takes into account the number of species present, as well as the relative abundance of each species. As species richness and evenness increase, so diversity increases.

$$D = 1 - \left(\frac{\Sigma\, n(n-1)}{N(N-1)}\right)$$

n = the total number of organisms of a particular species N = the total number of organisms of all species

The value of D ranges between 0 and 1. With this index, 1 represents infinite diversity and 0, no diversity.

Climate Resilient Technology (Y6)

Climate resilient technology here has been conceived in the form of those approaches or technologies, modern or traditional followed to increase the copping or adapting ability of existing form against the brunt of global warming and climate change. Here it has been measured by scoring different measures taken up by farmers of the study area to cope up with the changing climate and environment.

Ecosystem Output (Y7)

While farmers has been conceived here as a Holon of ecosystem, the ecological output indicate and assert the total volume of photosynthetic produced cut off system behabiour of that ecology. Here it has been measured in terms of kg/ha.

Table 5.3 List of Independent Variables

Sl. No.	Independent Variables	Notations
1.	Age	X1
2.	Education	X2
3.	Functional Education	X3
4.	Size of family members	X4
5.	Size of holding	X5
6.	Number of fragment	X6
7.	Average size of fragment	X7
8.	Mean distance between two fragments	X8
9.	Size of homestead land	X9
10.	Yield of Crop	X10
11.	Marketed surplus	X11
12.	Land under irrigation	X12
13.	Rainfed area	X13
14.	Cropping intensity	X14
15.	Total amount of seed used	X15
16.	Total hours of irrigation given	X16
17.	Total amount of chemical fertilizer applied	X17
18.	Count of livestock	X18
19.	On farm Income	X19
20.	Communication Variable	X20
21.	Consumption of coal, fuel wood, fire wood etc.(	X21
22.	Production of cowdung/poultry liter /goat/biogas slurry	X22

Age (X1)

In all societies, age is one of the most important determinants of social status and social role of the individual. In the present study, age of the respondent was measured on the basis of their chronological age at the time of investigation.

Education (X2)

Education is instrumental is building personality structure and helps in changing one's behavior in social life. Education may be conceptualized as the amount of formal schooling literacy acquired by the responded.

Functional Education (X3)

The respondents were asked how many days they received training regarding the technical knowhow of cultivation and also post harvest processing, value addition or any other activities etc. It is measured by adding total number of days they received training in each time.

Size of family members (X4)

It is the number of family members of individual farmers.

Size of holding (X5)

The amount of land owned by a person is an important parameter to access the economic status of the person in the society. Here it has been measured by Total area of land owned by the respondent divided by number of family members.

Number of fragment (X6)

Land fragments here have been perceived in terms of total number of entitlements under the total size of holding. For example – if one ha of land has got three geographical distributions than the number of fragmentation will be three. Here it has been measured by counting the Arithmetic count of pieces of the land (fragments) under the ownership of family.

Average size of fragment(X7)

Average or mean size of total number of fragments were taken into account.

Mean distance between two fragments(X8)

Mean distance between all fragments under the ownership of a family were measured.

Size of homestead land(X9)

Amount of land acquired by the home building and surrounding. This total area was measured in ha and divided by the family size.

Yield of Crop(X10)

Total yield of rice, vegetables and pulses were measured in kg/ha for one each family.

Marketed surplus(X11)

Total amount of produced in each month sold to the market by each respondent were measured in kg.

Land under irrigation (X12)

Total area under the ownership of one family covered under irrigated agriculture was measured in ha.

Rainfed area (X13)

Total area under the ownership of one family covered under rain fed agriculture was measured in ha.

Cropping intensity(X14)

It refers to the proportion of acreage annually put under different crops to the total cropped area as expressed in percentage. The cropping intensity is calculated by Cropping intensity (%) = (Gross Cropped Area/ Net Cropped Area) X 100

Total amount of seed used (X15)

Total amount of seed sowed to the field for rice, vegetable and pulses were measured in kg.

Total hours of irrigation given (X16)

Total hours of irrigation given in each season to rice, vegetables and pulses were measured in hours.

Total amount of chemical fertilizer applied (X17)

Total amount of chemical fertilizer applied in each season to rice, vegetables and pulses were measured in kg/ha.

Count of livestock (X18)

Arithmetic count cattle, poultry, goat possessed by each family were taken into account. Here it has been measured as Total number of cattle, poultry and goat divided by number of family member.

On farm Income (X19)

Total income by the respondent from farm in each month was measured in Rupees.

Communication Variable(X20)

It refers to the use of information source(s) both from outside and within the community for getting agricultural information. The score assigned as followed:

Tv/radio	Newspaper or any print media	Dealers or peer farmer	Experts from an Institution	calculation
A days in a week	B days in a week	C days in a week	D days in a week	A*12+B*12+C*12+D*12

Consumption of coal, fuel wood, fire wood etc.(X21)

Consumption of green sources of energy is a common habit in rural India. It reduces their cost of purchasing fuel mainly for culinary purpose from outside. Here it has been measured in terms of kg/ha of coal, fuel wood, fire wood, bamboo, dried and uneconomic part of crop used in each month by a family divided by number of item they use for the same.

Production of cowdung/poultry liter /goat/biogas slurry(X22)

Cow dung, poultry or goat liter, biogas slurry are an important domestic by product which subsequently improve the soil health and also crop quality and usher the possibilities of stewardship. Here it has been measured by taking into account the total amount of cow dung, poultry or goat liter, biogas slurry produced in a month for each in kg.

PREPARATION OF INTERVIEW SCHEDULE

On the basis of finding pilot study a preliminary schedule was formed with the help of literature and by the assistance of Chairman of Advisory Committee. The interview schedule consisted of three major parts according to the specific objective of the study.

PRE TESTING OF INTERVIEW SCHEDULE

It is the process of advance testing of the study design after the schedule has been prepared. The object of pretesting is to detect the discrepancies that have emerged and to remove them after necessary modification in the schedule. After conducting pretesting appropriate changes and modification of the interview schedule have been made. The individuals who responded in pretesting have been excluded in the final sample selected for the study.

TECHNIQUES OF FIELD DATA COLLECTION

The respondents were personally interviewed during durga puja vacation and summer vacation. The items were asked in local language, Bengali .The entire were donein the schedule by student investigator himself at the time of interview.

STATISTICAL TOOLS USED FOR ANALYSIS OF DATA

The statistical methods for analysis and interpretation of raw data were..

a. Range
b. Mean.
c. Standard deviation.
d. Coefficient of Variance.
e. Correlation of coefficient.
f. Step down regression analysis.
g. Dendogram analysis
h. Path analysis
i. Factor analysis
j. Canonical covariate analysis.

a. **Range:** In statistics, the range of a set of data is the difference between the largest and smallest values. It is expressed by maximum and minimum value of a set of data.

b. **Mean:** The mean is the arithmetic average and is the result obtained when the sum of the value of individual in the data is divided by the number of individuals in the data. Mean is and relatively stable measure of central tendency. The mean reflects and is by every score in the distribution.

When the data are expressed in a frequency distribution (grouped), the mean is calculated using the following formula

$$X = \frac{\sum_{1=1}^{N} fixi}{N}$$

Where,

x = Mean of the observation.

fi = Frequency of the class.

xi = Mid value of the class.

N = Total number of observation

c. **Standard deviation:** Standard deviation (SD) of a set of observation is the square root of the arithmetic mean of the squares of the deviations. The deviations being measured from the arithmetic mean of the distributions. It is commonly denoted by the symbol (Sigma). To measure the average deviation from the standard value of the data standard deviation is used. It is less affected by sampling en ors and is a more stable measure of dispersion.

The standard deviation of the data grouped in the form of frequency distribution is computed by using the following formula –

$$S.D. = \frac{\sum_{1=1}^{N} fixi}{N}\left\{\frac{\sum_{1=1}^{N} fixi}{N}\right\}$$

When,

d = Standard deviation

N = total No of observation in a particular coll.

X = value of observation in a particular cell

F = Frequency of observation

X = mean number of observation

I = any number (e.g. 1, 2, 3) denoting position

d. **Coefficient of Variance:** A measure of variation which is independent of the unit of measurement is proved by the coefficient of variation. Being unit free, this is useful For comparison of variability between different populations. The coefficient of variation is standard deviation expressed as percentage of the mean.

Coefficient of variation is measured by using the following formula —

$$CV = \frac{SD}{Mean} \times 100$$

e. **Coefficient of correlation:** When increase or decrease in one variety is accompanied by an increase or in another variety, the two are said to be correlated and Ute phenomenon is known as correlation. Correlation coefficient (r) is a measure of the relationship between two variables, which are at the interval or rational level of measurement and are linearly related. A Person product moment "r" is computed by the formula.

$$Rxy = \frac{N\dot{a}XY(\dot{a}X)(\dot{a}Y)}{\sqrt{[N\dot{a}X^2 - (\dot{a}X)^2][N\dot{a}Y^2 - (\dot{a}Y)^2]}}$$

When,

X and Y = Original scares in variables X and Y

N = Number of paired scores

å XY = Each X multiplied by its corresponding Y. then summed

å X = Sum of X scores

å X^2 = Each of X squared, then summed

(å $X)^2$ = Sum of X score squared

å Y = Sum of Y scores

å Y^2 = Each of Y squared, than summed

(å $Y)^2$ = Sum of Y score squared

The range of correlation coefficient is between 1 to +1. This perfect negative correlation +1 perfect positive correlation. A perfect correlation is, however, seldom achieved. A correlation Coefficient to be acceptable should be statistically significant. Otherwise, we say that no significant relationship exist between the variables.

f. **Step down regression analysis:** Stepwise regression is which provides a means of choosing independent variables that yield the best possible with the fewest independent variables. It permits the user to solve a sequence of one more multiple linear regression problems by stepwise application of the least square method. At each step in the analysis. a variable is added or removed which results in the greatest production in the error sum of squares.

g. **Dendrogram analysis:** A dendrogram (from Greek dendro "tree" and gramma "drawing") is a tree diagram frequently used to illustrate the arrangement of the clusters produced by hierarchical clustering based on Euclidean distances.

h. **Path Analysis:** Path Analysis is a straight forward extension of multiple regressions. Its aim is to provide estimations of the magnitude and significance of hypothesized causal connections between sets of variables. Path analysis is a statistical method with a structural model, but not measurement model. In other terms it is called as causal modeling, analysis of co variance structures and latent variable models. The reason for the name is that the techniques allow us to test theoretical propositions about cause and effect without manipulating variables. However, the causal in causal modeling refers to an assumption of the model rather than a property of the output or consequence of the technique. That is, people assume some variables are causally related and the test propositions about them using the technique. If the propositions are supported, it does not prove that the causal assumptions are correct.

i. **Factor Analysis:** Factor analysis is a very useful and popular method of multivariate research technique, mostly used in social and behavioral sciences. Factor analysis seeks to resolve a large set of measured variables in terms of relatively few categories, known as factors. This technique allows the researcher to group variables into factors (based on correlation between variables); the factors so derived may be treated as new variables (often termed as latent variables) and their value derived by summing the values of the original variables, which had been grouped into the factor. The meaning and name of such new variable is subjectively determined by the researcher. Since the factors happen to be linear combinations of data, the coordinates of each observation or variable is measured to obtain what are factor loadings. Such factor loading represent the correlation between the variable and the factor and are usually placed in a matrix of correlations of the variables and the factors. In the Factor Analysis the "Principle Component Method" was followed.

j. **Canonical covariate analysis:** In statistics, canonical covariate analysis (CCA) is a way of making sense of cross covariance matrices. If we have two vectors X = (Xi, ..Xa) and Y = (Yl, ... Ym) of random variables, and there are variables, then canonical correlation analysis will find linear combinations of the Xi and Yj whish have maximum correlation with each other. Virtually all of the commonly encountered parametric tests of significance can be treated as special cases of canonical correlation analysis, which is the general procedure for investigating the relationships between two sets of variables. The method was first introduced by Harold Hotelling in I936.

Given two column vectors X = (x1.....xa)1 and Y = (y1...ym)1 of random variables with finite second moments, one may define the cross covariance to be the n m matrix whose (i.j) entry is the convarience CON(r,y). In the covariance CON(r,y). In practice, we would estimate the covariance matrix based on sampled data from X and Y (i.e. from a pair of data matrices).

Canonical correlation analysis seeks vectors a′ and b′ such that the random variables a1X and b′Y maximize the correlation p = corr(a′X,b′Y). The random variables U =a′X and V = b′Y are the first pair of canonical variables. Then one seeks vectors maximizing the same correlation subject to the constraint that they are to be uncorrelated with the first pair of canonical variables; this gives the second pair of canonical variables. This procedure may be continued up to min{m, n} times.

6
CHAPTER

EMPIRICAL STUDIES : PART A

Participatory Analysis And Delineation Onbaseline Characters

Pra (Participatory Rural Appraisal) Exercise and Revelation

The social ecology of this study area is undergoing moderate to fast transformation with simmering impact on social life, institutional functions, natural resource management and livelihood therein. Participatory analysis of a social ecology has so far been complex, multidimensional, collateral and concomitant. The space, time, decision and flow analysis are generally followed to present the process of characterization and pattering changes across the social ecology. The space analysis, depicted by social map, offers the splendid opportunity to study the socio geographical distribution, ecological bondage and functional interdependency amongst and between different constituent biosocial and bio physical components.

The time line and time trend analysis delineate and describes the dent and pattern of changes over and across space and time. The flow analysis on the other hand offers the unique vision on the networking of different organization, institutions and functions therewith.

Participatory analysis and Stewardship: Stewardship function offers a spatial and temporal distribution, which again can be delved into community learning process, nicely reflected through different participatory mapping, analysis and networking. The participatory process, mostly exercised through PRA practices, has well been understood in the form of visualization, experiential learning, and empowerment, integrating competency and community conclusion and dialogues. So, the present study reveals not only participatory tools and techniques but also the process of instilling spirits of empowerment and method of mutual sharing and synchronization, both of methods and motivation, techniques and learning, philosophy and planning.

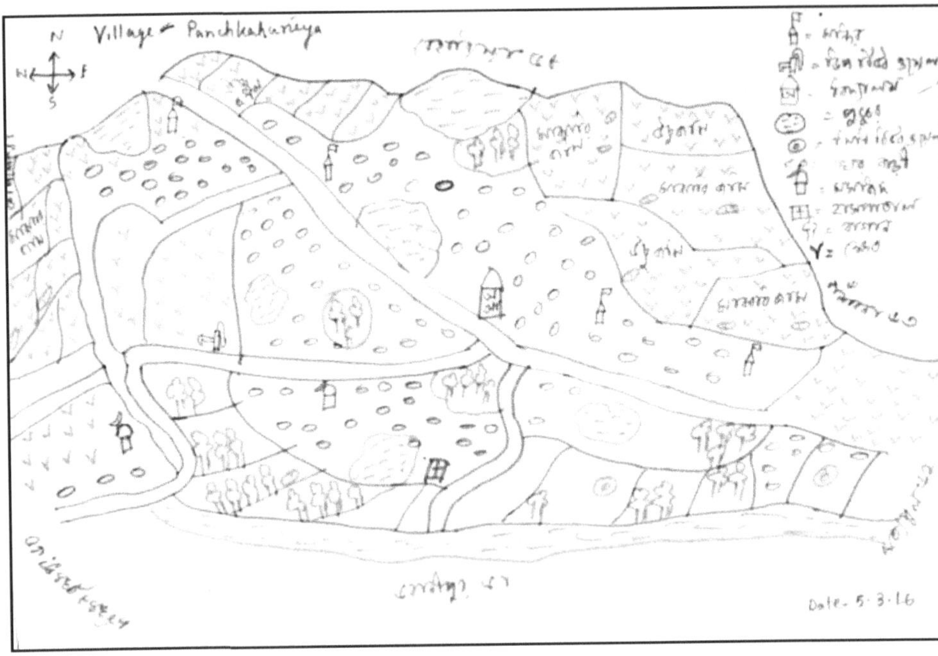

Figure: 6.A.1: Delineation 1 : Social Map of the Villages

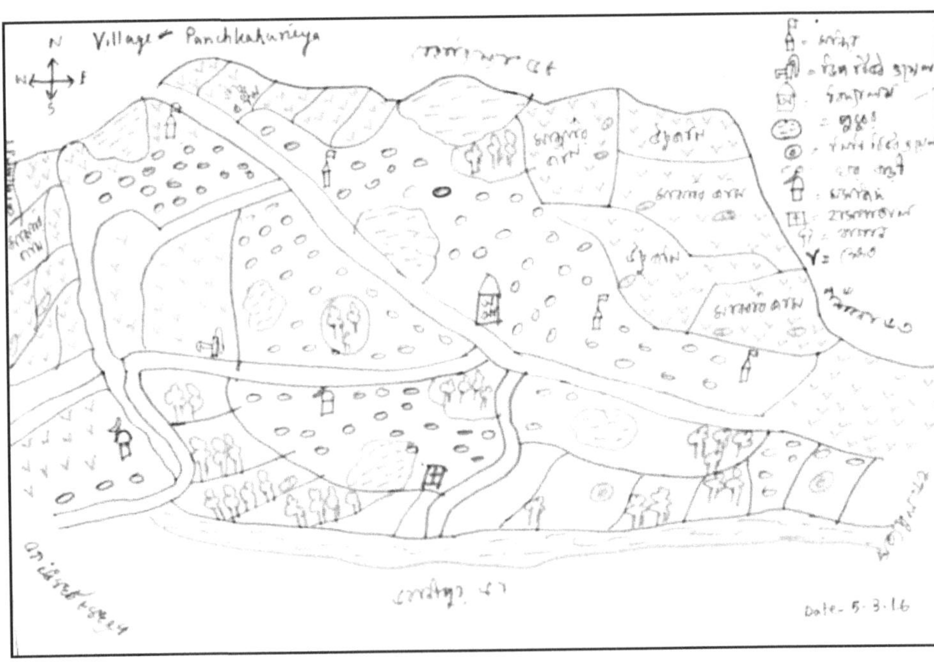

Plate 1 Social Map of Panchkahaniya

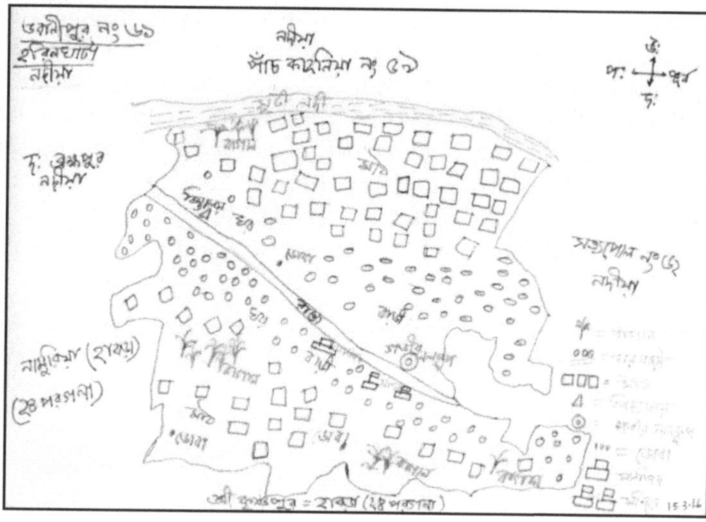

Plate 2 Social Map of Bhabanipur

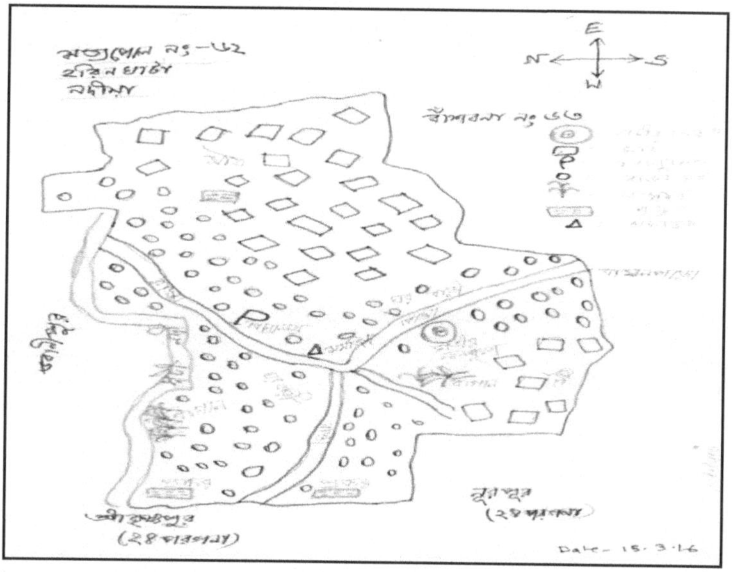

Plate 3 Social Map of Satyapul

DELINEATION 1 PRESENTS THE SOCIAL MAP OF VILLAGES

The PRA exercise was conducted on 15th March and 3rd March, 2016, by the following villagers. Kenaram Ghosh, Amit Ghosh for village Panchkahaniya, Imran naji molla, Sarif ali Mondal, Kuddum Mondal for village Satyapul, Suvankar Mondal, Shambhu Biswas, Prashanta Mondal for village Bhabanipur. The Social map here has depicted that the villagers are mostly hailing from small and marginal holding groups. The biodiversities are well distributed across the villages and settlement patterns are clustered evenly across the villages. The unique features as depicted by the social map are as follows; There are good numbers of lentic water bodies, mainly used for retting of jute, partly used for fishery, and used as sources of drinking water for livestock. The villages are encircled with the Suti river and the water used for vegetable cultivation. The village has got the following cropping pattern: Rice-Jute-Cauliflower, Tomato, Sesame, local Arum, banana, pointed guard, Tuberose which lays the foundation for their agrarian economy.

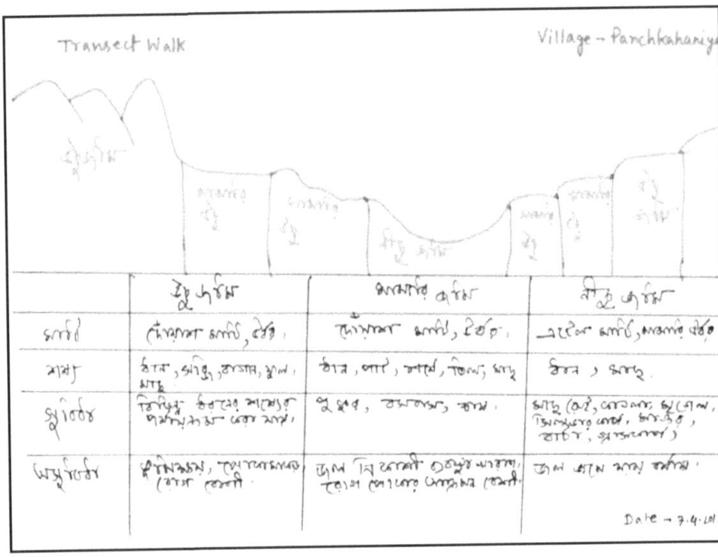

Plate 4-Transect Walk of Panchkahaniya

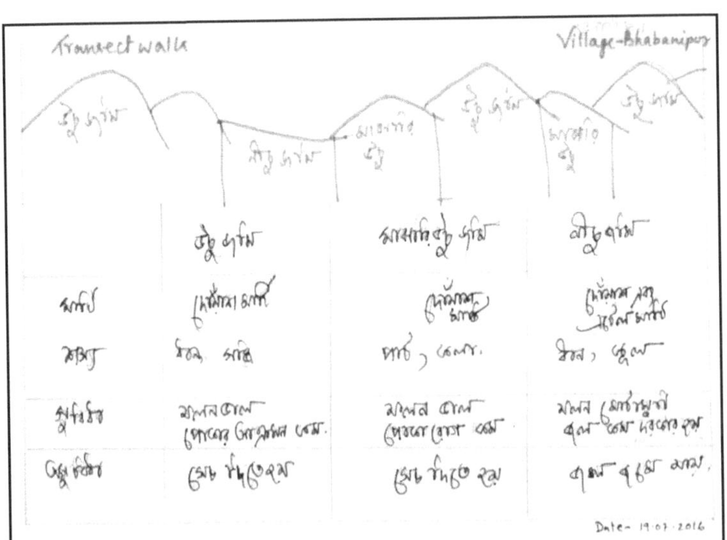

Plate 5-Transect Walk of Bhabanipur

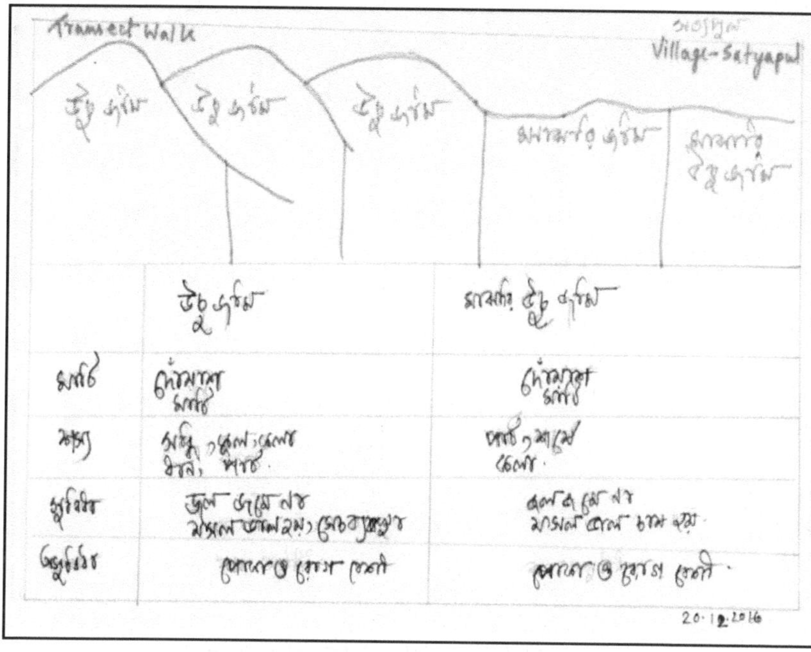

Plate 6-Transect Walk of Satyapul

Figure : 6.A.2.Delineation 2 : Transect walk of the Villages

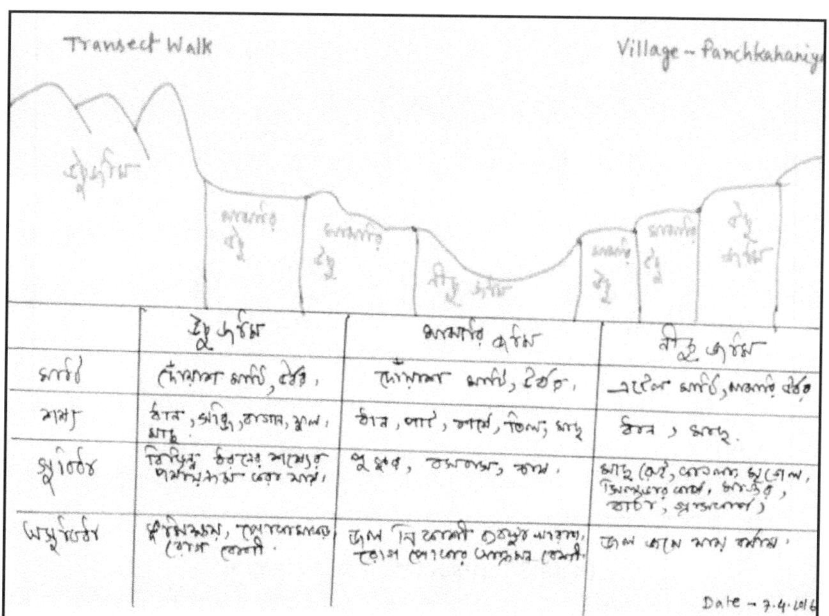

Plate 4 Transect Walk of Panchkahaniya

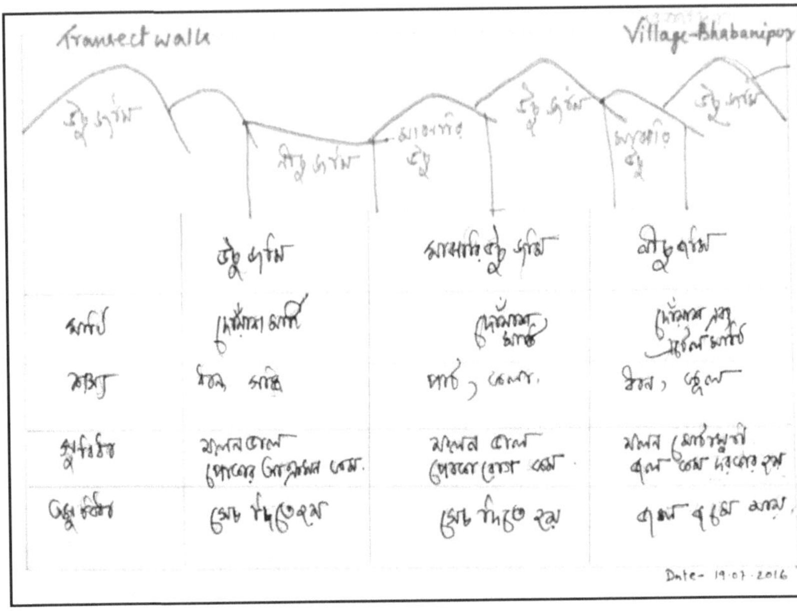

Plate 5 Transect Walk of Bhabanipur

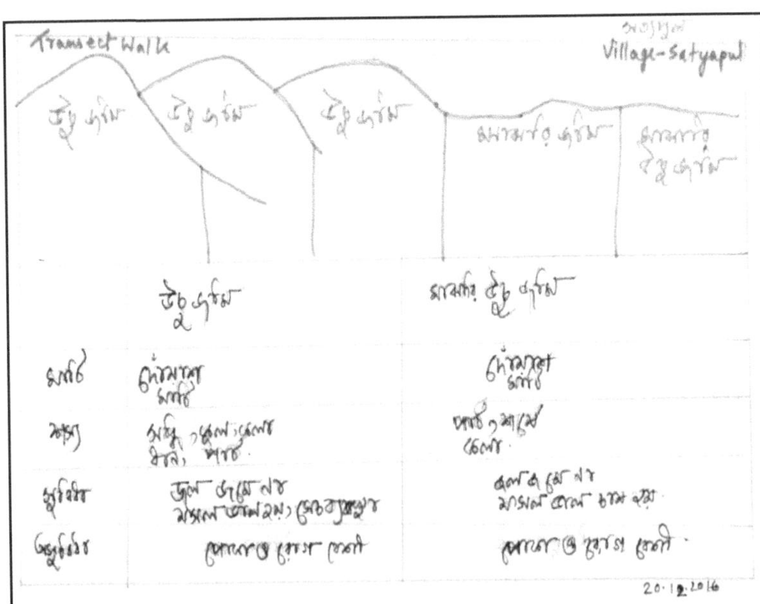

Plate 6 Transect Walk of Satyapul

DELINEATION 2:

The transect walk has been conducted by forming a heterogeneous group of villagers led by an opinion leader of the villages and the walk was conducted across the village on 7.04.2016, 19.7.2016, 20.12.2016. The topography of the villages includes high, medium and low lands. In addition to rice, jute, pulse, vegetables and orchard crops, fish biodiversities are very predominant for the low lying areas. While orchard is very popular option for high land, field crops for medium land, summer rice and fishing are very common for low lying area. Both inundation and soil erosion are discernable problems for the agro ecosystem associated with the indiscriminate use of pesticides and fertilizer as well.

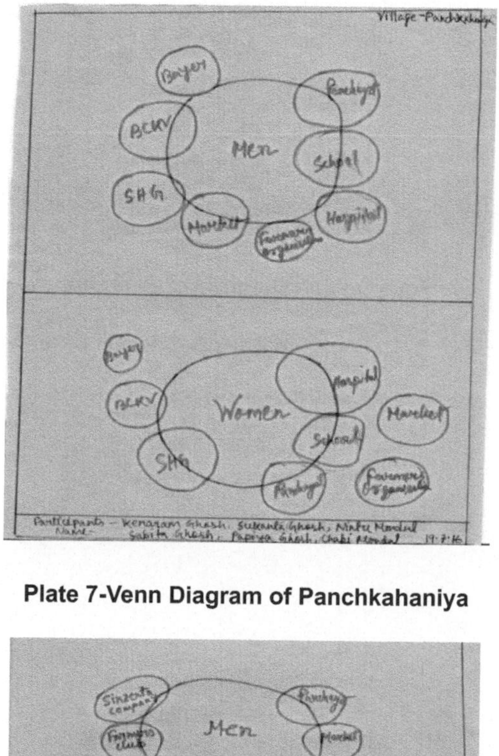

Plate 7-Venn Diagram of Panchkahaniya

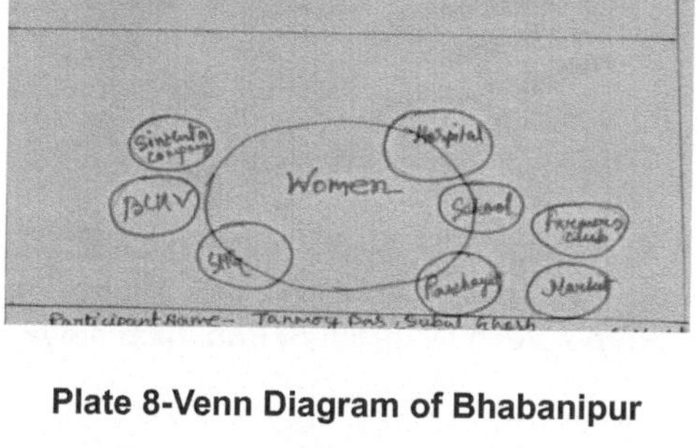

Plate 8-Venn Diagram of Bhabanipur

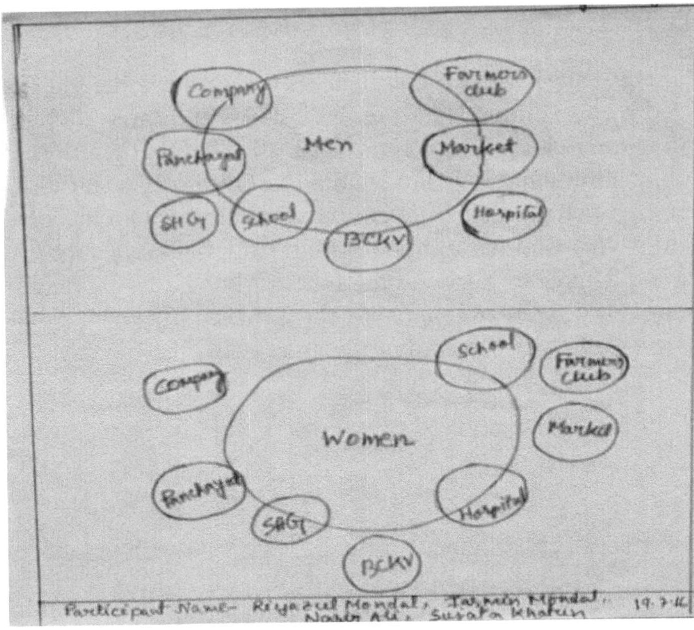

Plate 9-Venn Diagram of Satyapul

Figure: 6.A.3. Delineation 3 Venn diagram of the villages

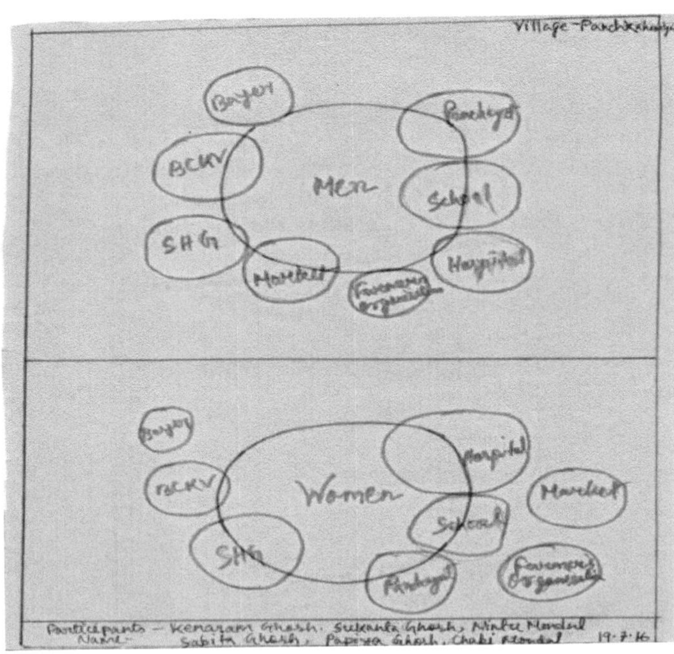

Plate 7 Venn Diagram of Panchkahaniya

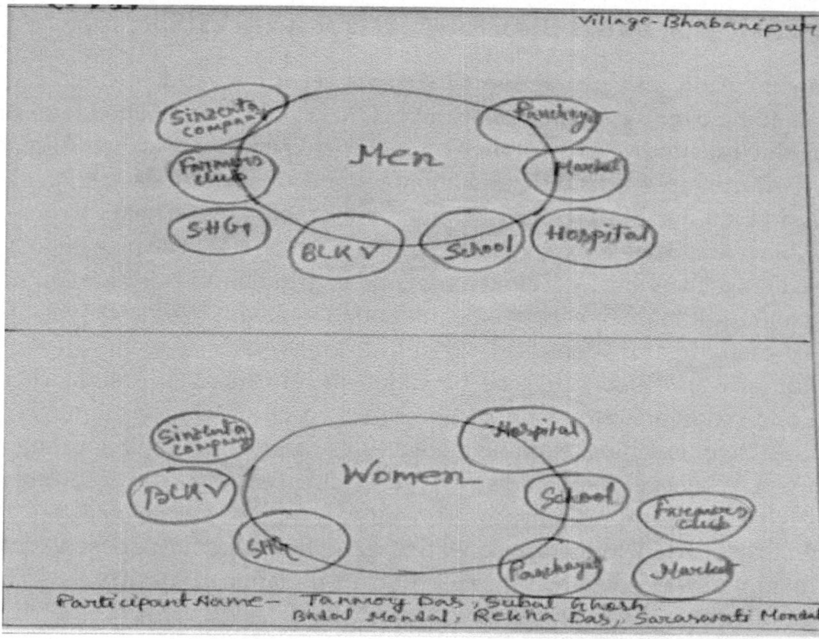

Plate 8 Venn Diagram of Bhabanipur

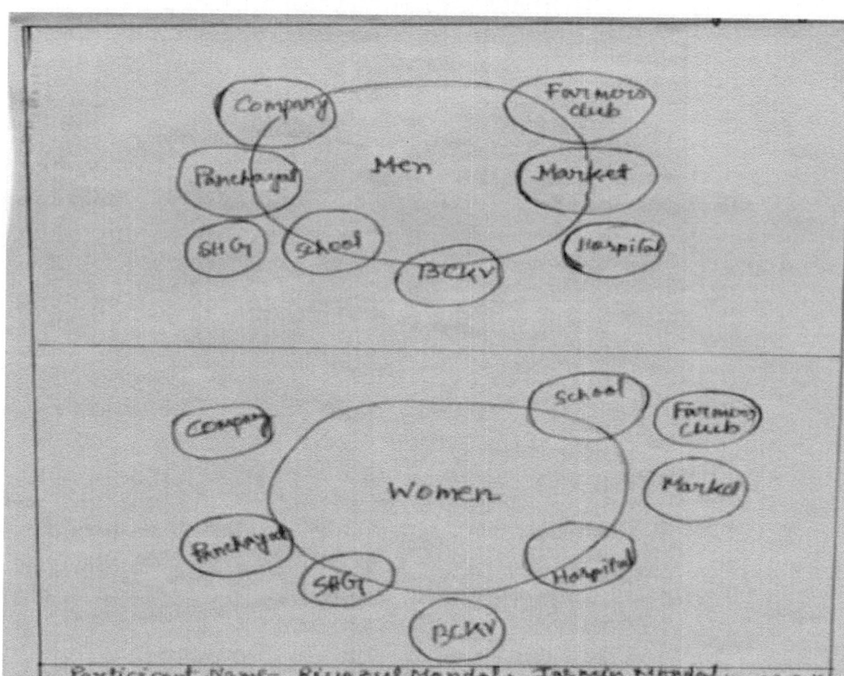

Plate 9 Venn Diagram of Satyapul

DELINEATION 3: VENN DIAGRAM (FLOW ANALYSIS)

For conservation and protection of natural resources and to have a better understanding over the institutional support that can be extrapolated for exerting stewardship function Venn diagram is an important participatory tool. To generate better stewardship the leading stakeholders and organizations are Panchayet, School, Hospital, and Market. SHG, Farmers organizations, Agrochemical companies (IFFCo, Sinzenta, Bayer etc), agricultural university (BCKV). The programmes like afforestation, cataloguing local biodiversity and registration and creation of nutrition garden and local tourism can well be integrated by involving Panchayet, School, SHG, Farmers organizations and agricultural university as well. Mobilization of community and their active participation in controlling as well as checking of soil and biodiversity erosion can be achieved by creating community gene bank, collective mulching, conjunctive water use and community auditing (Pani Panchayet). Whenever livelihoods keep generating out of stewardship functions, classical conservation approaches can pave the way for transformation of, e.g., Forest conservation into rural tourism, means an up gradation of resource conservation into a community stewardship, tourism, means an up gradation of resource conservation into a community stewardship

Time line Analysis

TIME LINE — PANCHKAHANIYA

YEAR	IMPORTANT EVENT
1958	SEVERE DROUGHT
1963	HAILSTONE ALMOST FOR 7 DAYS
1964	RADIO CAME
1977	FLOOD OCCURED
1978	DEEPTUBE WELL INSTALLED.
1987	TV CAME.
2004	BIOGAS INSTALLED
2009	BRICK ROAD CONSTRUCTED
2011	PUCCA ROAD CONSTRUCTED
	Participant Name- Astocharan Ghosh, Nimaichandra Ghosh, kartik Ghosh Bibwanath Ghosh.

Date-6.3.17.

Plate 10 Time line Analysis of Panchkahaniya

TIME LINE BHABANIPUR

YEAR	IMPORTANT EVENT
1958	DROUGHT OCCURED
1960	FLOOD OCCURED
1965	BICYCLE CAME
1970	DEEP TUBE WELL INSTALLED
1980	MOTOR BIKE CAME
1980	TV CAME
2001	BRICK ROAD CONSTRUCTED
2008	BIOGAS INSTALLED.
2010	PUCCA-ROAD CONSTRUCTED.

6. 3. 17.

Plate 11 Time line Analysis of Bhabanipur

TIME LINE SATYAPUL

YEAR	IMPORTANT EVENT
1958	DROUGHT OCCURED
1960	BICYCLE CAME
1965	HEAVY FLOOD
1970	FLOOD
1975	DEEP TUBE WELL INSTALLED
1977	FLOOD
1985	TV CAME
1990	MOTOR BIKE CAME
2004	BIOGAS INSTALLED
2009	BRICK ROAD CONSTRUCTED
2011	PUCCA ROAD CONSTRUCTED

DATE— 6. 3. 17.

Plate 12 Time line Analysis of Satyapu

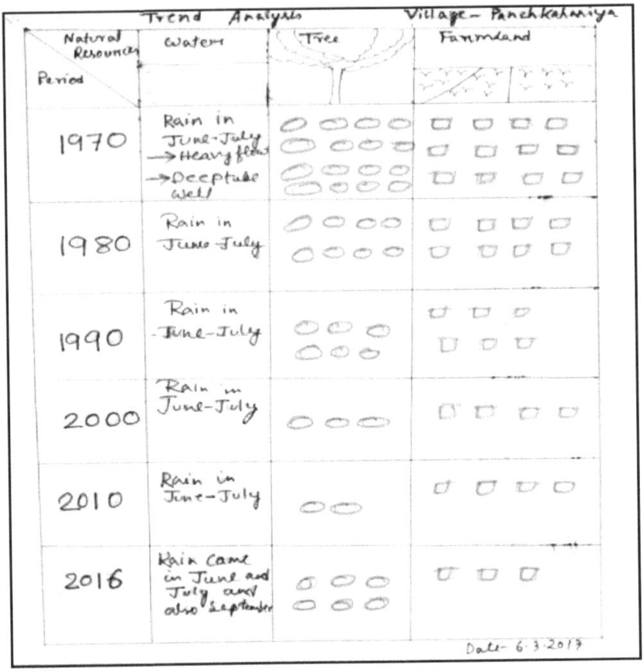

Plate 13 -Trend Analysis of Panchkahaniya

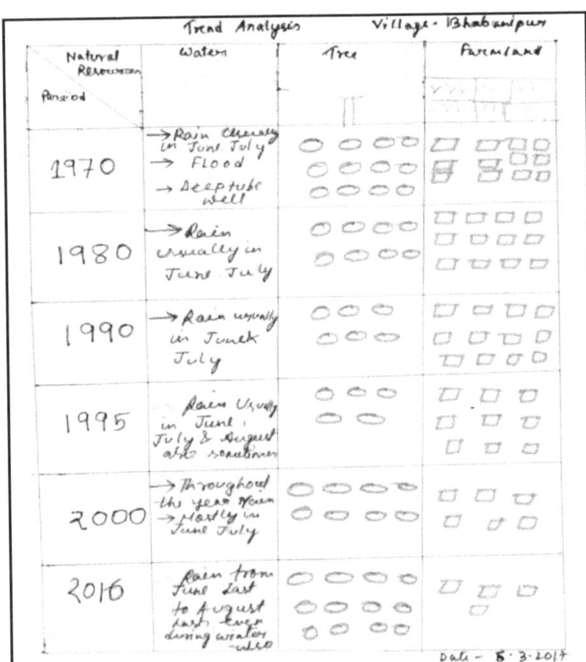

Plate 14-Trend Analysis of Bhabanipur

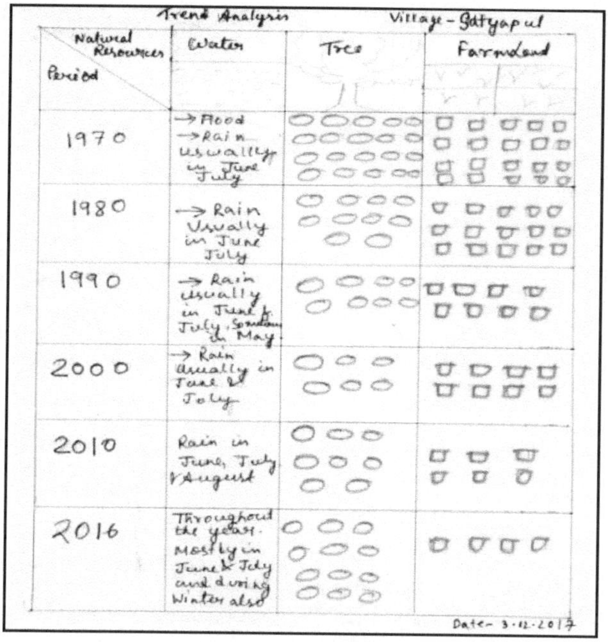

Plate 15 -Trend Analysis of Satyapul

Figure: 6.A.4. Delineation 4: Time Analysis

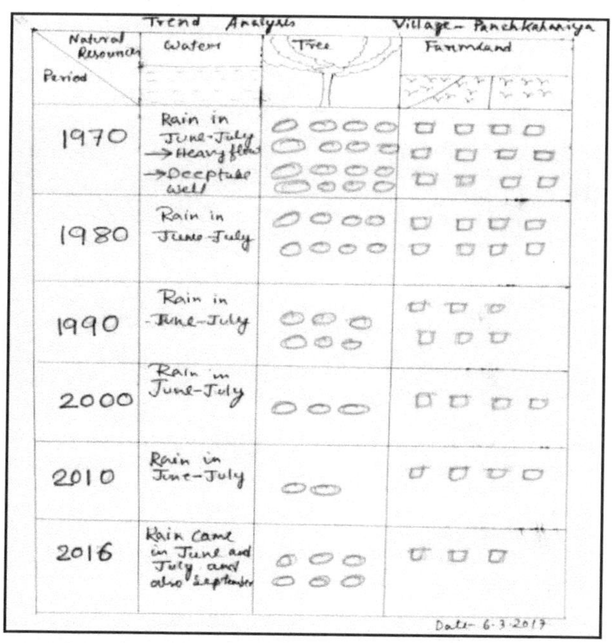

Plate 13 Trend Analysis of Panchkahaniya

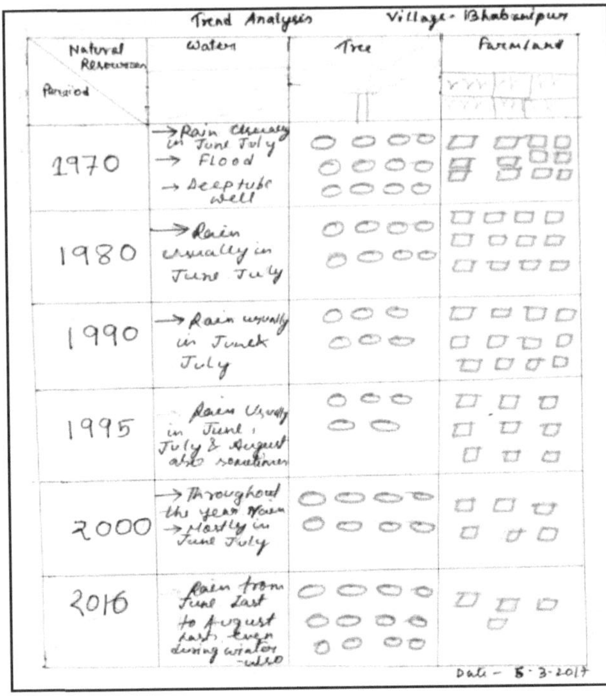

Plate 14 Trend Analysis of Bhabanipur

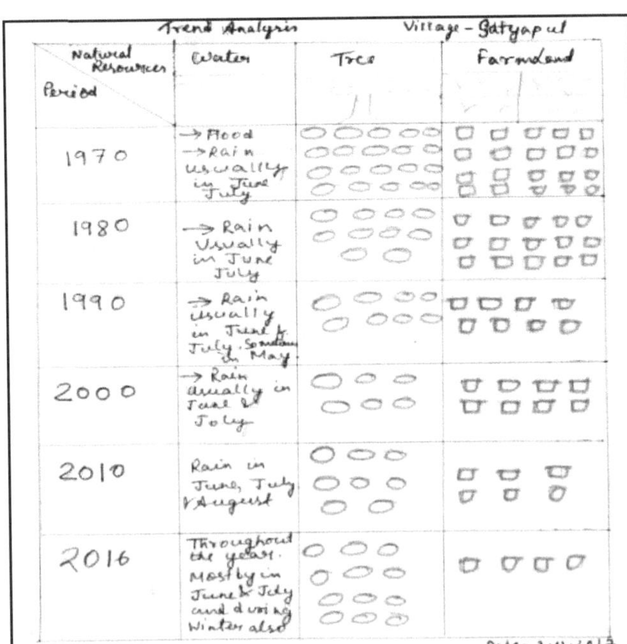

Plate 15 Trend Analysis of Satyapul

DELINEATION 4: TIME ANALYSIS

Participatory analysis of time and its chronological perception reflected through changed pattern and changed direction have leaded the foundation for time analysis. The analysis was conducted by the villagers named Astocharan Ghosh, Nimaichandra ghosh, Kartik Ghosh, Biswanath Ghosh on 6.03.17, 3.02.17 for the 3 villages.

Time analysis has got one time lime and another is Time trend. The time line connects the events over the year, time trend delineates the pattern of change over decades.

Time trend analysis presents that fact elucidate is that "between 1970 to 2016" the farm lands available per family have been sustainably reduced. This is a stress on land resource generated by teeming population. So far has biodiversity of tree and plantation have also substantially reduced until 2010 but the indication of retrieval, may be due to social forestry programme is well discernible.

Another trend has been well conspicuous about monsoon rain behavior "between 1970 to 2016". The community perception suggest the entry of monsoon has been delayed by one to two weeks even a month and the monsoon has also been extended up to last week of September to 1st week of October. When there is a clear swing of rain for the cohort of post monsoon period. The worst effected aspect of farm enterprise has been seed bed preparation for early vegetable crops as well as harvesting of autumn paddy.

The time line depicts that the community have identified deep tube well installation, flood, and entry of TV, Biogas and brick road construction as the marker events.

Introduction of biogas can be traced out as one of the important stewardship functions to conserve the both soil and energy, to offer much required ecological deliverables to the community.

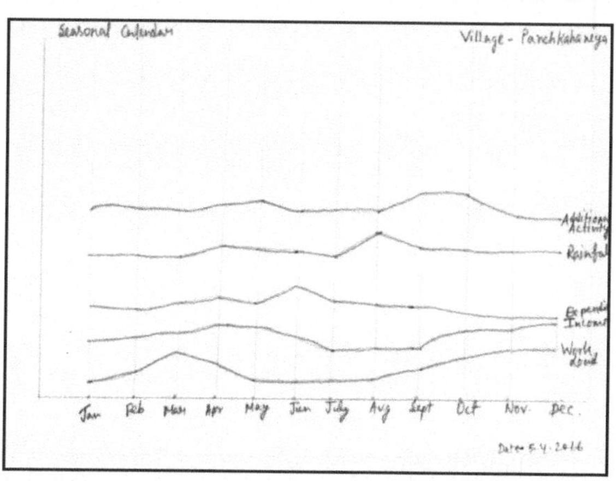

Plate 16-Seasonal Calendar of Panchkahaniya

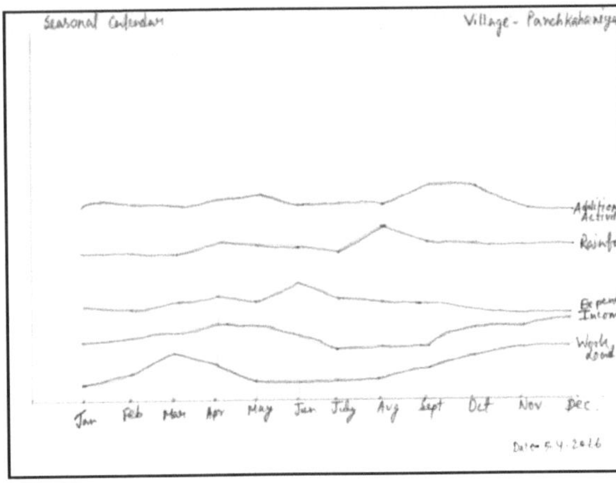

Plate 16 Seasonal Calendar of Panchkahaniya

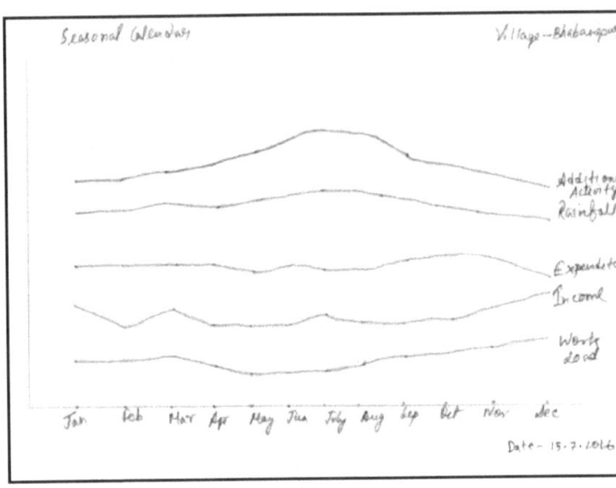

Plate 17-Seasonal Calendar of Bhabanipur

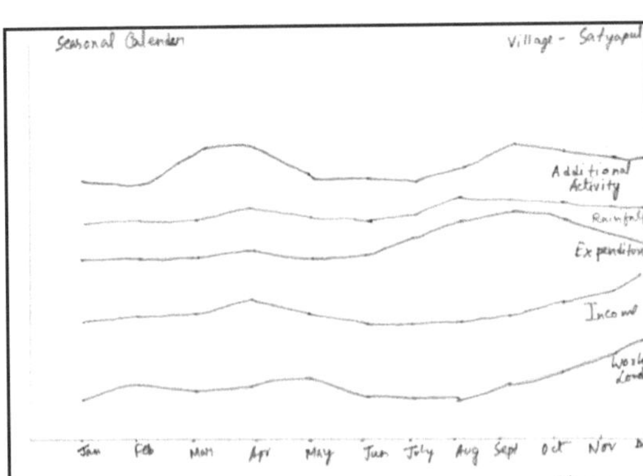

Plate 18-Seasonal Calendar of Satyapul

Figure: 6.A.5.

DELINEATION 5: SEASONAL CALENDAR OF VILLAGES

Seasonal calendar offers the researcher to have a glance at the variability of different livelihood functions across the months. The expenditure shoots up during June, July, while income of farm families increases during April May, may be associated with MGNREGA activities, intercultural operations for vegetables and harvesting of paddy. While rainfall is the maximum during August, income, expenditure as well as work load decline. However, off farm activities are important job opportunity for the rural people which got a steady distribution over the months.

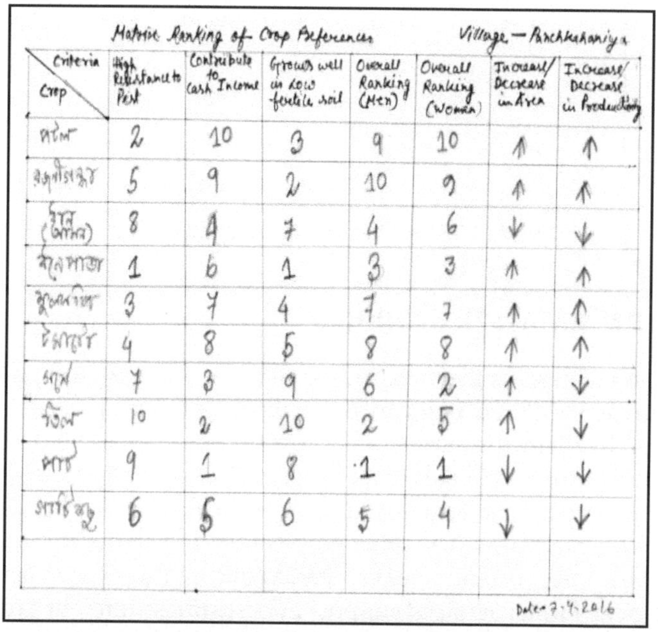

Matrix Ranking of Crop Preferences — Village – Panchkhaniya

Criteria / Crop	High Resistance to Pest	Contribute to Cash Income	Grows well in low fertile soil	Overall Ranking (Men)	Overall Ranking (Women)	Increase/ Decrease in Area	Increase/ Decrease in Productivity
পটল	2	10	3	9	10	↑	↑
[illegible]	5	9	2	10	9	↑	↑
ধান (আমন)	8	4	7	4	6	↓	↓
[illegible]	1	6	1	3	3	↑	↑
[illegible]	3	7	4	7	7	↑	↑
[illegible]	4	8	5	8	8	↑	↑
[illegible]	7	3	9	6	2	↑	↓
[illegible]	10	2	10	2	5	↑	↓
পাট	9	1	8	1	1	↓	↓
[illegible]	6	5	6	5	4	↓	↓

Date- 7.4.2016

Figure: 6.A.6.Delineation 6: Matrix ranking of villages

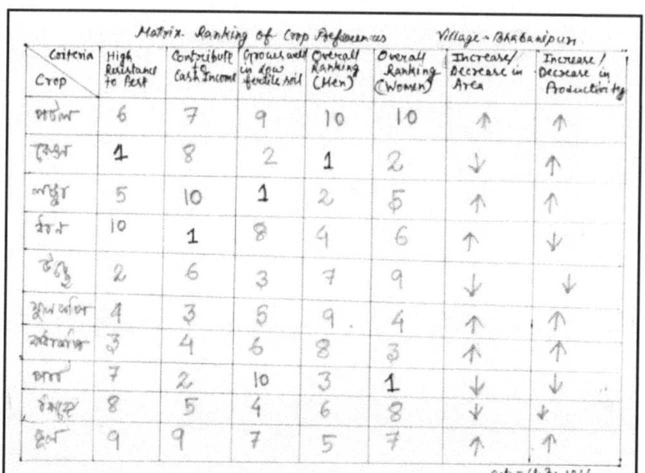

Matrix. Ranking of Crop Preferences — Village - Bhabanipur

Criteria / Crop	High Resistance to Pest	Contribute to Cash Income	Grows well in low fertile soil	Overall Ranking (Men)	Overall Ranking (Women)	Increase/ Decrease in Area	Increase / Decrease in Productivity
পটল	6	7	9	10	10	↑	↑
[illegible]	1	8	2	1	2	↓	↑
[illegible]	5	10	1	2	5	↑	↑
[illegible]	10	1	8	4	6	↑	↓
[illegible]	2	6	3	7	9	↓	↓
[illegible]	4	3	5	9	4	↑	↑
[illegible]	3	4	6	8	3	↑	↑
পাট	7	2	10	3	1	↓	↓
[illegible]	8	5	4	6	8	↓	↓
[illegible]	9	9	7	5	7	↑	↑

Date - 11.7.2016

Plate 19 Matrix Ranking of Panchkahaniya

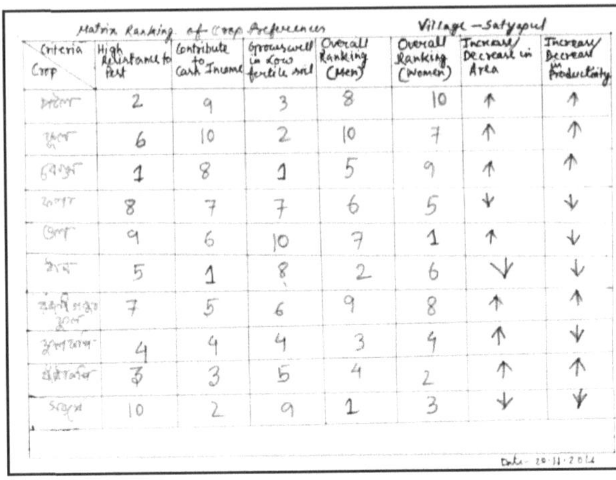
Matrix Ranking of Crop Preferences Village – Satyapul

Criteria / Crop	High Resistance to Pest	Contribute to Cash Income	Grows well in Low fertile soil	Overall Ranking (Men)	Overall Ranking (Women)	Increase/ Decrease in Area	Increase/ Decrease in Productivity
[illegible]	2	9	3	8	10	↑	↑
[illegible]	6	10	2	10	7	↑	↑
[illegible]	1	8	1	5	9	↑	↑
[illegible]	8	7	7	6	5	↓	↓
[illegible]	9	6	10	7	1	↑	↓
[illegible]	5	1	8	2	6	↓	↓
[illegible]	7	5	6	9	8	↑	↑
[illegible]	4	4	4	3	4	↑	↓
[illegible]	3	3	5	4	2	↑	↑
[illegible]	10	2	9	1	3	↓	↓

Date - 20.11.2014

Plate 20 Matrix Ranking of Bhabanipur

DELINEATION 6 MATRIX RANKING

Delineation 6 presents the matrix ranking to elucidate choices and preferences by focus group over different enterprises, livelihood or events based on community criteria derived by them, the matrix ranking, an important tool for decision analysis here shows that for these villages crop, vegetables and floriculture enterprises like tuberose, pointed guard and tomato have earned clear choices, made by the farmers as the profitable enterprises. The other crop enterprises in the fray are local arum, jute, sesame, mustard, coriander leaves, paddy and cauliflower. When community decisions or peer farmer choices support and sustains different crops, having a support from operating market and technical backup, the crop biodiversity is certain to be maintained, just to cherish an diverse biogenomics, managed and mentored by the community people.

7

CHAPTER

EMPIRICAL STUDIES : PART B

Table 6.B.1: Descriptive statistics of Village 1 (Panchkahaniya)

Variable	Range		Mean	SD	CV(%)
	Max.	Min.			
Age (X1)	63.70	37.80	49.09	7.37	0.15
Education (X2)	12.00	1.00	6.08	3.22	0.53
Functional Education (X3)	44.00	3.00	11.68	7.59	0.65
Size of family members (X4)	7.00	4.00	5.16	0.87	0.17
Size of holding (X5)	0.27	0.04	0.12	0.05	0.43
Number of fragment (X6)	18.00	5.00	9.62	3.37	0.35
Average size of fragment(X7)	0.13	0.03	0.06	0.02	0.31
Mean distance between two fragments(X8)	0.98	0.44	0.71	0.13	0.18
Size of homestead land(X9)	15.00	8.00	12.18	1.93	0.16
Yield of Crop(X10)	63150.00	43500.00	54668.25	5364.15	0.10
Marketed surplus(X11)	59265.00	38439.75	49065.98	5299.42	0.11
Land under irrigation (X12)	0.93	0.20	0.53	0.18	0.35
Rainfed area (X13)	0.33	0.00	0.09	0.09	0.99
Cropping intensity(X14)	300.00	223.08	261.23	25.96	0.10
Total amount of seed used (X15)	120.45	79.20	98.68	10.67	0.11
Total hours of irrigation given (X16)	11.00	7.50	9.67	0.94	0.10
Total amount of chemical fertilizer applied (X17)	1087.50	881.25	986.70	57.50	0.06
Count of livestock (X18)	6.50	1.14	3.22	1.14	0.35

Variable	Range		Mean	SD	CV(%)
	Max.	Min.			
On farm Income (X19)	7500.00	4000.00	5734.00	909.27	0.16
Communication Variable(X20)	237.00	45.00	74.82	26.44	0.35
Consumption of coal, fuel wood, fire wood etc.(X21)	261.25	90.00	139.23	42.44	0.30
Production of cowdung/ poultry liter /goat/biogas slurry(X22)	125.50	8.00	51.69	30.70	0.59
Application of organic manure (Y1)	15000.00	3750.00	8100.00	3400.84	0.42
Rice stubble height left in the Field (Y2)	7.00	3.00	5.04	1.37	0.27
Crop residue left in the field (Y3)	23295.00	3045.00	6180.75	2653.80	0.43
Soil conservation level (Y4)	0.60	0.06	0.21	0.13	0.60
Crop Biodiversity conservation (Y5)	0.44	0.27	0.29	0.03	0.09
Climate resilient technology (Y6)	6.00	3.00	4.46	1.18	0.26
Ecosystem output (Y7)	70620.00	50291.25	60849.00	5364.50	0.09

VILLAGE 1

The distribution pattern of variable, **Age (X1),** depicts that the minimum age of the respondents is 37.8 and maximum is 63.7. The mean age is 49.088 with standard deviation 7.365879. The coefficient of variance (CV) is 0.150055 per cent which indicates that the distribution pattern of the variable is consistent.

The distribution pattern of variable, **Education (X2),** shows that the minimum standard of the respondents is 1 and the maximum is 12. The mean education is 6.08 with S.D of 3.22. The coefficient of variance is 0.529 per cent which indicates distribution of the variables is fairly consistent.

The distribution pattern of variable, **Functional Education (X3),** shows that the minimum standard of the respondents is 3 and the maximum is 44. The mean of education is 11.68 with S.D of 7.59. The coefficient of variance is 0.650 per cent which indicates distribution of the variables is fairly consistent

The distribution pattern of variable, **Family size (X4),** depicts that the maximum family size of the respondents is 7 and the minimum is 4. The mean of size of the family is 5.16 with standard deviation of 0.866. The coefficient of variance is 0.168 per cent which indicates the distribution of variable is fairly consistent.

The distribution pattern of variable, **Size of holding (X5),** depicts that the maximum size of holding is 0.267 and the minimum is 0.038. The mean size of holding is 0.123 with standard deviation of 0.05. The coefficient of variance is 0.429 per cent which indicates the distribution of variable is fairly consistent.

The distribution pattern of variable, **Number of fragment (X6),** depicts that the maximum number of fragment is 18 and the minimum is 5. The mean number of fragment is 9.62 with standard deviation of 3.368. The coefficient of variance is 0.350 per cent which indicates the distribution of variable is fairly consistent.

The distribution pattern of variable, **Average size of fragment (X7),** depicts that the maximum size of fragment is 0.133 and the minimum is 0.03. The mean size of fragment is 0.064 with standard deviation of 0.314. The coefficient of variance is 0.350 per cent which indicates the distribution of variable is fairly consistent.

The distribution pattern of variable, **Mean distance between two fragments (X8),** depicts that the maximum distance between two fragments is 0.975 and the minimum is 0.443. The mean distance between two fragments is 0.708 with standard deviation of 0.127. The coefficient of variance is 0.180 per cent which indicates the distribution of variable is fairly consistent.

The distribution pattern of variable, **Size of homestead land (X9),** depicts that the maximum size is 15 and the minimum is 8. The mean size of homestead land is 12.18 with standard deviation of 1.934. The coefficient of variance is 0.159 per cent which indicates the distribution of variable is fairly consistent.

The distribution pattern of variable, **Yield of Crop (X10),** depicts that the maximum yield is 63150 and the minimum is 43500. The mean yield of crop is 54668.25 with standard deviation of 5364.15. The coefficient of variance is 0.098 per cent which indicates the distribution of variable is fairly consistent.

The distribution pattern of variable, **Marketed surplus (X11),** depicts that the maximum marketed surplus is 59265 and the minimum is 38439.75. The mean ofmarketed surplus is 49065.98 with standard deviation of 5299.41. The coefficient of variance is 0.108 per cent which indicates the distribution of variable is fairly consistent.

The distribution pattern of variable, **Land under irrigation (X12),** depicts that the maximum land under irrigation is 0.933 and the minimum is 0.2. The mean land under irrigation is 0.525 with standard deviation of 0.182. The coefficient of variance is 0.346 per cent which indicates the distribution of variable is fairly consistent.

The distribution pattern of variable, **Rainfed area (X13**), depicts that the maximum rainfed area is 0.33 and the minimum is 0. The mean of rainfed area is 0.093 with standard deviation of 0.092. The coefficient of variance is 0.98 per cent which indicates the distribution of variable is fairly consistent.

The distribution pattern of variable, **Cropping intensity (X14),** depicts that the maximum cropping intensityis 300 and the minimum is 223.07. The mean of cropping intensity is 261.23 with standard deviation of 25.958. The coefficient of variance is 0.099 per cent which indicates the distribution of variable is fairly consistent.

The distribution pattern of variable, **Total amount of seed used (X15**), depicts that the maximum amount of seed used is 120.45 and the minimum is 79.2. The mean amount of seed used is 98.68 with standard deviation of 10.67. The coefficient of variance is 0.108 per cent which indicates the distribution of variable is fairly consistent.

The distribution pattern of variable, **Total hours of irrigation given (X16),** depicts that the maximum hours of irrigation given is 11 and the minimum is 7.5. The mean for total hours of irrigation given is 9.67 with standard deviation of 0.940. The coefficient of variance is 0.097 per cent which indicates the distribution of variable is fairly consistent.

The distribution pattern of variable, **Total amount of chemical fertilizer applied (X17)**, depicts that the maximum amount of chemical fertilizer applied is 1087.5 and the minimum is 881.25. The mean of amount of chemical fertilizer applied is 0.525 with standard deviation of 57.50. The coefficient of variance is 0.059 per cent which indicates the distribution of variable is fairly consistent.

The distribution pattern of variable, **Count of livestock (X18),** depicts that the maximum count of livestock is 6.5 and the minimum is 1.142. The mean count of livestock is 3.219 with standard deviation of 1.136. The coefficient of variance is 0.353 per cent which indicates the distribution of variable is fairly consistent.

The distribution pattern of variable, **On farm income (X19),** depicts that the maximum on farm income is 7500 and the minimum is 4000. The mean of on farm income is 5734 with standard deviation of 909.27. The coefficient of variance is 0.159 per cent which indicates the distribution of variable is fairly consistent.

The distribution pattern of **Communication variable (X20),** depicts that the maximum communication variable is 237 and the minimum is 45. The mean of communication variable is 74.82 with standard deviation of 26.436. The coefficient of variance is 0.353 per cent which indicates the distribution of variable is fairly consistent.

The distribution pattern of variable **Consumption of coal, fuel wood, fire wood etc. (X21),** depicts that the maximum consumption is 261.25 and the minimum is 90. The mean consumption of coal, fuel wood, fire wood etc. is 139.225 with standard deviation of 42.441. The coefficient of variance is 0.305 per cent which indicates the distribution of variable is fairly consistent.

The distribution pattern of variable, **Production of cowdung/ poultry liter / goat/biogas slurry (X22),** depicts that the maximum production is 125.5 and the minimum is 8. The mean production of cowdung/ poultry liter /goat/biogas slurry is 51.69 with standard deviation of 30.70. The coefficient of variance is 0.594 per cent which indicates the distribution of variable is fairly consistent.

The distribution pattern of variable, **Application of organic manure (Y1),** shows that the maximum organic manure applied by the respondent is 15000 while the minimum is 3750. The mean of application of organic manure is 8100 with a standard deviation of 3432.69. The coefficient of variance is 0.420 percent which indicates that the distribution of the variable is quite consistent.

The distribution pattern of variable, **Rice stubble height left in the field (Y2),** shows that the maximum stubble height left in the field is 7 while the minimum is 3. The mean of rice stubble height left in the field is 5.04 with a standard deviation of

1.370. The coefficient of variance is 0.272 percent which indicates that the distribution of the variable is quite consistent.

The distribution pattern of variable, **Crop residue left in the field (Y3),** shows that the maximum crop residue left by the respondent is 23295 while the minimum is 3045. The mean of crop residue left in the field is 6180.75 with a standard deviation of 2653.80. The coefficient of variance is 0.429 percent which indicates that the distribution of the variable is quite consistent.

The distribution pattern of variable, **Soil conservation level (Y4),** shows that the maximum level of soil conservation is 0.6 while the minimum is 0.06. The mean level of soil conservation is 0.21 with a standard deviation of 0.127. The coefficient of variance is 0.602 percent which indicates that the distribution of the variable is quite consistent.

The distribution pattern of variable, **Crop biodiversity conservation (Y5),** shows that the maximum crop biodiversity conservation followed by the respondent is 0.444 while the minimum is 0.2667. The mean of crop biodiversity conservation is 0.29 with a standard deviation of 0.0259. The coefficient of variance is 0.089 percent which indicates that the distribution of the variable is quite consistent.

The distribution pattern of variable, **Climate resilient technology (Y6),** shows that the maximum climate resilient technology followed by the respondent is 6 while the minimum is 3. The mean of climate resilient technology followed is 4.46 with a standard deviation of 1.182. The coefficient of variance is 0.265 percent which indicates that the distribution of the variable is quite consistent.

The distribution pattern of variable, **Ecosystem output (Y7),** shows that the energy output is 70620 while the minimum is 50291.25. The mean of energy output is 60849 with a standard deviation of 5364.50. The coefficient of variance is 0.0881 percent which indicates that the distribution of the variable is quite consistent.

Table 6.B.2 : Descriptive Statistics of Village 2 (Bhabanipur)

Variable	Range		Mean	SD	CV(%)
	Max.	Min.			
Age (X1)	60.00	20.00	39.01	8.23	0.21
Education (X2)	12.00	5.00	9.48	1.67	0.18
Functional Education (X3)	29.00	0.00	9.58	5.96	0.62
Size of family members (X4)	12.00	3.00	6.62	2.58	0.39
Size of holding (X5)	0.20	0.04	0.10	0.04	0.40
Number of fragment (X6)	7.00	4.00	5.04	0.90	0.18
Average size of fragment(X7)	0.21	0.08	0.12	0.03	0.27
Mean distance between two fragments(X8)	1.00	0.24	0.59	0.17	0.29
Size of homestead land(X9)	0.53	0.00	0.10	0.07	0.74

Variable	Range		Mean	SD	CV(%)
	Max.	Min.			
Yield of Crop(X10)	33450.00	12480.00	27197.25	3968.44	0.15
Marketed surplus(X11)	4133.00	1349.68	3157.68	524.52	0.17
Land under irrigation (X12)	0.71	0.20	0.49	0.12	0.24
Rainfed area (X13)	0.26	0.00	0.10	0.08	0.81
Cropping intensity(X14)	303.03	245.79	277.48	13.88	0.05
Total amount of seed used (X15)	105.30	64.05	82.85	9.44	0.11
Total hours of irrigation given (X16)	13.00	6.50	9.77	1.42	0.15
Total amount of chemical fertilizer applied (X17)	900.00	487.50	703.35	101.07	0.14
Count of livestock (X18)	4.00	0.10	1.43	0.80	0.56
On farm Income (X19)	8500.00	4000.00	6668.00	984.46	0.15
Communication Variable(X20)	129.00	33.00	93.60	23.21	0.25
Consumption of coal, fuel wood, fire wood etc.(X21)	68.00	35.00	44.45	7.39	0.17
Production of cowdung/ poultry liter /goat/biogas slurry(X22)	110.50	6.00	55.39	20.53	0.37
Application of organic manure (Y1)	3000.00	750.00	1831.50	499.36	0.27
Rice stubble height left in the Field (Y2)	6.00	2.00	3.78	1.11	0.29
Crop residue left in the field (Y3)	6045.00	3435.00	4865.70	684.73	0.14
Soil conservation level (Y4)	0.23	0.04	0.08	0.03	0.43
Crop Biodiversity conservation (Y5)	0.31	0.20	0.25	0.02	0.09
Climate resilient technology (Y6)	5.00	2.00	3.64	0.83	0.23
Ecosystemoutput (Y7)	38737.50	17542.50	32062.95	4246.09	0.13

The distribution pattern of variable, **Age (X1),** depicts that the minimum age of the respondents is 20 and maximum is 60. The mean age is 39.012with standard deviation 8.226. The coefficient of variance (CV) is 0.211per cent which indicates that the distribution pattern of the variable is consistent.

The distribution pattern of variable, **Education (X2),** shows that the minimum standard of the respondents is 5 and the maximum is 12. The mean of education is 9.48 with S.D of 1.669. The coefficient of variance is 0.176 per cent which indicates distribution of the variables is fairly consistent.

The distribution pattern of variable, **Functional education (X3),** shows that the minimum standard of the respondents is 0 and the maximum is 29. The mean of functional education is 9.58 with S.D of 5.96. The coefficient of variance is 0.622 per cent which indicates distribution of the variables is fairly consistent

The distribution pattern of variable, **Family size (X4),** depicts that the maximum family size of the respondents is 12 and the minimum is 3. The mean of size of the family is 6.62 with standard deviation of 2.578. The coefficient of variance is 0.386 per cent which indicates the distribution of variable is fairly consistent.

The distribution pattern of variable, **Size of holding (X5),** depicts that the maximum size of holding is 0.198 and the minimum is 0.038. The mean size of holding is 0.103 with standard deviation of 0.04. The coefficient of variance is 0.406 per cent which indicates the distribution of variable is fairly consistent.

The distribution pattern of variable, **Number of fragment (X6),** depicts that the maximum number of fragment is 7 and the minimum is 4. The mean number of fragment is 5.04 with standard deviation of 0.902. The coefficient of variance is 0.179 per cent which indicates the distribution of variable is fairly consistent.

The distribution pattern of variable, **Average size of fragment (X7),** depicts that the maximum size of fragment is 0.215 and the minimum is 0.075. The mean of average size of fragment is 0.122 with standard deviation of 0.033. The coefficient of variance is 0.274 per cent which indicates the distribution of variable is fairly consistent.

The distribution pattern of variable, **Mean distance between two fragments (X8)**, depicts that the maximum distance between two fragments is 1 and the minimum is 0.244. The mean distance between two fragments is 0.591with standard deviation of 0.169. The coefficient of variance is 0.285 per cent which indicates the distribution of variable is fairly consistent.

The distribution pattern of variable, **Size of homestead land (X9),** depicts that the maximum size is 0.533 and the minimum is 0.005. The mean size of homestead land is 0.098with standard deviation of 0.072. The coefficient of variance is 0.737 per cent which indicates the distribution of variable is fairly consistent.

The distribution pattern of variable, **Yield of crop (X10),** depicts that the maximum yield is 33450 and the minimum is 12480. The mean yield of crop is 27197.25 with standard deviation of 3968.43. The coefficient of variance is 0.146 per cent which indicates the distribution of variable is fairly consistent.

The distribution pattern of variable, **Marketed surplus (X11),** depicts that the maximum marketed surplus is 4133 and the minimum is 1349.68. The mean of marketed surplus is 3157.675 with standard deviation of 524.52. The coefficient of variance is 0.166 per cent which indicates the distribution of variable is fairly consistent.

The distribution pattern of variable, **Land under irrigation (X12),** depicts that the maximum land under irrigation is 0.706 and the minimum is 0.2. The mean land under irrigation is 0.489 with standard deviation of 0.116. The coefficient of variance is 0.240 per cent which indicates the distribution of variable is fairly consistent.

The distribution pattern of variable, **Rainfed area (X13)**, depicts that the maximum rainfed area is 0.262 and the minimum is 0.004. The mean of rainfed area is 0.103with standard deviation of 0.0833. The coefficient of variance is 0.083 per cent which indicates the distribution of variable is fairly consistent.

The distribution pattern of variable, **Cropping intensity (X14),** depicts that the maximum cropping intensity is 303.03 and the minimum is 245.79. The mean of cropping intensity is 277.48 with standard deviation of 13.877. The coefficient of variance is 0.050 per cent which indicates the distribution of variable is fairly consistent.

The distribution pattern of variable, **Total amount of seed used (X15**), depicts that the maximum amount of seed used is 105.3 and the minimum is 64.05. The mean amount of seed used is 82.84 with standard deviation of 9.439. The coefficient of variance is 0.114 per cent which indicates the distribution of variable is fairly consistent.

The distribution pattern of variable, **Total hours of irrigation given (X16),** depicts that the maximum hours of irrigation given is 13 and the minimum is 6.5. The mean for total hours of irrigation given is 9.774with standard deviation of 1.420. The coefficient of variance is 0.145 per cent which indicates the distribution of variable is fairly consistent.

The distribution pattern of variable, **Total amount of chemical fertilizer applied (X17),** depicts that the maximum amount of chemical fertilizer applied is 900 and the minimum is 487.50. The mean of amount of chemical fertilizer applied is 703.35with standard deviation of 101.07. The coefficient of variance is 0.144 per cent which indicates the distribution of variable is fairly consistent.

The distribution pattern of variable, **Count of livestock (X18),** depicts that the maximum count of livestock is 4 and the minimum is 0.10. The mean count of livestock is 1.430 with standard deviation of 0.799. The coefficient of variance is 0.559 per cent which indicates the distribution of variable is fairly consistent.

The distribution pattern of variable, **On farm income (X19),** depicts that the maximum on farm income is 8500 and the minimum is 4000. The mean of on farm income is 6668 with standard deviation of 984.45. The coefficient of variance is 0.148 per cent which indicates the distribution of variable is fairly consistent.

The distribution pattern of **Communication variable (X20),** depicts that the maximum communication variable is 129 and the minimum is 33. The mean of communication variable is 93.6 with standard deviation of 23.214. The coefficient of variance is 0.248 per cent which indicates the distribution of variable is fairly consistent.

The distribution pattern of variable, **Consumption of coal, fuel wood, fire wood etc. (X21),** depicts that the maximum consumption is 68 and the minimum is 35. The mean consumption of coal, fuel wood, fire wood etc. is 44.45 with standard deviation of 7.38. The coefficient of variance is 0.166 per cent which indicates the distribution of variable is fairly consistent.

The distribution pattern of variable, **Production of cowdung/ poultry liter / goat/biogas slurry (X22),** depicts that the maximum production is 110.5 and the minimum is 6. The mean production of cowdung/ poultry liter /goat/biogas slurry is 55.39 with standard deviation of 20.52. The coefficient of variance is 0.371 per cent which indicates the distribution of variable is fairly consistent.

The distribution pattern of variable, **Application of organic manure (Y1),** shows that the maximum organic manure applied by the respondent is 3000 while the minimum is 750. The mean of application of organic manure is 1831.5 with a standard deviation of 499.357. The coefficient of variance is 0.273 percent which indicates that the distribution of the variable is quite consistent.

The distribution pattern of variable, **Rice stubble height left in the field (Y2),** shows that the maximum stubble height left in the field is 6 while the minimum is 2. The mean of rice stubble height left in the field is 3.78 with a standard deviation of 1.11. The coefficient of variance is 0.294 percent which indicates that the distribution of the variable is quite consistent.

The distribution pattern of variable, **Crop residue left in the field (Y3),** shows that the maximum crop residue left by the respondent is 6045 while the minimum is 3435. The mean of crop residue left in the field is 4865.7 with a standard deviation of 684.3. The coefficient of variance is 0.141 percent which indicates that the distribution of the variable is quite consistent.

The distribution pattern of variable, **Soil conservation level (Y4),** shows that the maximum level of soil conservation is 0.232 while the minimum is 0.037. The mean level of soil conservation is 0.076 with a standard deviation of 0.032. The coefficient of variance is 0.427 percent which indicates that the distribution of the variable is quite consistent.

The distribution pattern of variable, **Crop biodiversity conservation (Y5),** shows that the maximum crop biodiversity conservation followed by the respondent is 0.311 while the minimum is 0.2. The mean of crop biodiversity conservation is 0.254 with a standard deviation of 0.0228. The coefficient of variance is 0.090 percent which indicates that the distribution of the variable is quite consistent.

The distribution pattern of variable, **Climate resilient technology (Y6),** shows that the maximum climate resilient technology followed by the respondent is 5 while the minimum is 2. The mean of climate resilient technology followed is 3.64 with a standard deviation of 0.827. The coefficient of variance is 0.227 percent which indicates that the distribution of the variable is quite consistent.

The distribution pattern of variable, **Ecosystem output (Y7),** shows that the energy output is 38737.5 while the minimum is 17542.5. The mean of energy output is 32062.95 with a standard deviation of 4246.093. The coefficient of variance is 0.13243 percent which indicates that the distribution of the variable is quite consistent.

Table 6.B.3 : Descriptive Statistics of Village 3 (Satyapul)

Variable	Range		Mean	SD	CV(%)
	Max	Min			
Age (X1)	66.00	27.00	44.95	9.85	0.22
Education (X2)	17.00	4.00	9.50	2.62	0.28
Functional Education (X3)	21.00	1.00	9.56	4.72	0.49

Variable	Range		Mean	SD	CV(%)
	Max	Min			
Size of family members (X4)	6.00	2.00	3.94	0.98	0.25
Size of holding (X5)	0.36	0.02	0.16	0.07	0.42
Number of fragment (X6)	8.00	1.00	4.32	1.68	0.39
Average size of fragment(X7)	0.31	0.07	0.14	0.05	0.38
Mean distance between two fragments(X8)	1.40	0.00	0.40	0.23	0.56
Size of homestead land(X9)	0.15	0.03	0.07	0.03	0.42
Yield of Crop(X10)	62550.00	28200.00	41807.25	7424.56	0.18
Marketed surplus(X11)	58860.00	23190.00	37443.90	6987.14	0.19
Land under irrigation (X12)	1.00	0.10	0.45	0.19	0.42
Rainfed area (X13)	0.67	0.00	0.14	0.13	0.97
Cropping intensity(X14)	300.00	212.50	267.13	16.41	0.06
Total amount of seed used (X15)	150.45	67.65	101.68	18.54	0.18
Total hours of irrigation given (X16)	22.00	3.50	11.18	4.45	0.40
Total amount of chemical fertilizer applied (X17)	2175.00	885.00	1255.35	266.88	0.21
Count of livestock (X18)	6.67	0.00	1.62	1.41	0.87
On farm Income (X19)	12000.00	2500.00	5200.00	2106.84	0.41
Communication Variable(X20)	270.00	45.00	115.80	51.62	0.45
Consumption of coal, fuel wood, fire wood etc.(X21)	28.75	3.00	20.66	5.20	0.25
Production of cowdung/ poultry liter /goat/biogas slurry(X22)	33.00	0.00	13.05	8.74	0.67
Application of organic manure (Y1)	1500.00	150.00	759.00	339.96	0.45
Rice stubble height left in the Field (Y2)	8.00	4.00	5.66	1.02	0.18
Crop residue left in the field (Y3)	5647.50	2265.00	4443.60	859.43	0.19
Soil conservation level (Y4)	0.90	0.08	0.27	0.15	0.56
Crop Biodiversity conservation (Y5)	0.33	0.24	0.27	0.03	0.10
Climate resilient technology (Y6)	6.00	1.00	3.70	1.50	0.41
Ecosystem output (Y7)	66345.00	33510.00	46250.85	7240.79	0.16

VILLAGE 3

The distribution pattern of variable, **Age (X1),** depicts that the minimum age of the respondents is 27 and maximum is 66. The mean age is 44.95with standard deviation 9.855. The coefficient of variance (CV) is 0.219per cent which indicates that the distribution pattern of the variable is consistent.

The distribution pattern of variable, **Education (X2)**, shows that the minimum standard of the respondents is 4and the maximum is 17. The mean of education is 9.5 with S.D of 2.620. The coefficient of variance is 0.276 per cent which indicates distribution of the variables is fairly consistent.

The distribution pattern of variable, **Functional education (X3),** shows that the minimum standard of the respondents is 1 and the maximum is 21. The mean of functional education is 9.5 with S.D of 2.620. The coefficient of variance is 0.276 per cent which indicates distribution of the variables is fairly consistent

The distribution pattern of variable, **Family size (X4),** depicts that the maximum family size of the respondents is 6 and the minimum is 2. The mean of size of the family is 3.94 with standard deviation of 0.978. The coefficient of variance is 0.248 per cent which indicates the distribution of variable is fairly consistent.

The distribution pattern of variable, **Size of holding (X5),** depicts that the maximum size of holding is 0.356 and the minimum is 0.022. The mean size of holding is 0.156 with standard deviation of 0.065. The coefficient of variance is 0.419 per cent which indicates the distribution of variable is fairly consistent.

The distribution pattern of variable, **Number of fragment (X6),** depicts that the maximum number of fragment is 8 and the minimum is 1. The mean number of fragment is 4.32 with standard deviation of 1.684. The coefficient of variance is 0.390 per cent which indicates the distribution of variable is fairly consistent.

The distribution pattern of variable, **Average size of fragment (X7),** depicts that the maximum size of fragment is 0.311 and the minimum is 0.006. The mean of average size of fragment is 0.144 with standard deviation of 0.054. The coefficient of variance is 0.380 per cent which indicates the distribution of variable is fairly consistent.

The distribution pattern of variable, **Mean distance between two fragments (X8),** depicts that the maximum distance between two fragments is 1.4 and the minimum is 0. The mean distance between two fragments is 0.404with standard deviation of 0.227. The coefficient of variance is 0.562 per cent which indicates the distribution of variable is fairly consistent.

The distribution pattern of variable, **Size of homestead land (X9),** depicts that the maximum size is 0.147 and the minimum is 0.027. The mean size of homestead land is 0.075 with standard deviation of 0.032. The coefficient of variance is 0.424 per cent which indicates the distribution of variable is fairly consistent.

The distribution pattern of variable, **Yield of crop (X10),** depicts that the maximum yield is 62550 and the minimum is 28200. The mean yield of crop is 41807.25 with standard deviation of 7424.559. The coefficient of variance is 0.178 per cent which indicates the distribution of variable is fairly consistent.

The distribution pattern of variable, **Marketed surplus (X11),** depicts that the maximum marketed surplus is 58860 and the minimum is 23190. The mean of marketed surplus is 37443.9with standard deviation of 6987.135. The coefficient of variance is 0.187 per cent which indicates the distribution of variable is fairly consistent.

The distribution pattern of variable, **Land under irrigation (X12),** depicts that the maximum land under irrigation is 1 and the minimum is 0.10. The mean land under irrigation is 0.446 with standard deviation of 0.186. The coefficient of variance is 0.418 per cent which indicates the distribution of variable is fairly consistent.

The distribution pattern of variable, **Rainfed area (X13**), depicts that the maximum rainfed area is 0.667 and the minimum is 0. The mean of rainfed area is 0.137with standard deviation of 0.133. The coefficient of variance is 0.968 per cent which indicates the distribution of variable is fairly consistent.

The distribution pattern of variable, **Cropping intensity (X14),** depicts that the maximum cropping intensityis 300 and the minimum is 212.50. The mean of cropping intensity is 26.129 with standard deviation of 16.413. The coefficient of variance is 0.061 per cent which indicates the distribution of variable is fairly consistent.

The distribution pattern of variable, **Total amount of seed used (X15**), depicts that the maximum amount of seed used is 150.45 and the minimum is 67.65. The mean amount of seed used is 101.675 with standard deviation of 18.54. The coefficient of variance is 0.183 per cent which indicates the distribution of variable is fairly consistent.

The distribution pattern of variable, **Total hours of irrigation given (X16),** depicts that the maximum hours of irrigation given is 22 and the minimum is 3.5. The mean for total hours of irrigation given is 11.175 with standard deviation of 4.454. The coefficient of variance is 0.399 per cent which indicates the distribution of variable is fairly consistent.

The distribution pattern of variable, **Total amount of chemical fertilizer applied (X17),** depicts that the maximum amount of chemical fertilizer applied is 2175 and the minimum is 885. The mean of amount of chemical fertilizer applied is 1255.35with standard deviation of 266.88. The coefficient of variance is 0.213 per cent which indicates the distribution of variable is fairly consistent.

The distribution pattern of variable, **Count of livestock (X18),** depicts that the maximum count of livestock is 6.667 and the minimum is 0. The mean count of livestock is 1.621with standard deviation of 1.410. The coefficient of variance is 0.870 per cent which indicates the distribution of variable is fairly consistent.

The distribution pattern of variable, **On farm income (X19),** depicts that the maximum on farm income is 8500 and the minimum is 4000. The mean of on farm income is 6668 with standard deviation of 984.45. The coefficient of variance is 0.148 per cent which indicates the distribution of variable is fairly consistent.

The distribution pattern of **Communication variable (X20),** depicts that the maximum communication variable is 270 and the minimum is 45. The mean of communication variable is 115.8 with standard deviation of 51.62. The coefficient of variance is 0.446 per cent which indicates the distribution of variable is fairly consistent.

The distribution pattern of variable,**Consumption of coal, fuel wood, fire wood etc. (X21),** depicts that the maximum consumption is 28.75 and the minimum is 3. The mean consumption of coal, fuel wood, fire wood etc. is 20.66 with standard deviation of 5.202. The coefficient of variance is 0.670 per cent which indicates the distribution of variable is fairly consistent.

The distribution pattern of variable, **Production of cowdung/ poultry liter /goat/ biogas slurry (X22),** depicts that the maximum production is 33 and the minimum is 0. The mean production of cowdung/ poultry liter /goat/biogas slurry is 13.052 with standard deviation of 8.74. The coefficient of variance is 0.371 per cent which indicates the distribution of variable is fairly consistent.

The distribution pattern of variable, **Application of organic manure (Y1),** shows that the maximum organic manure applied by the respondent is 1500 while the minimum is 150. The mean of application of organic manure is 759 with a standard deviation of 339.96. The coefficient of variance is 0.448 percent which indicates that the distribution of the variable is quite consistent.

The distribution pattern of variable, **Rice stubble height left in the field (Y2),** shows that the maximum stubble height left in the field is 8 while the minimum is 4. The mean of rice stubble height left in the field is 5.66 with a standard deviation of 1.022. The coefficient of variance is 0.181 percent which indicates that the distribution of the variable is quite consistent.

The distribution pattern of variable, **Crop residue left in the field (Y3),** shows that the maximum crop residue left by the respondent is 5647.5 while the minimum is 2265. The mean of crop residue left in the field is 4443.6 with a standard deviation of 859.43. The coefficient of variance is 0.193 percent which indicates that the distribution of the variable is quite consistent.

The distribution pattern of variable, **Soil conservation level (Y4),** shows that the maximum level of soil conservation is 0.9 while the minimum is 0.075. The mean level of soil conservation is 0.268 with a standard deviation of 0.150. The coefficient of variance is 0.561 percent which indicates that the distribution of the variable is quite consistent.

The distribution pattern of variable, **Crop biodiversity conservation (Y5),** shows that the maximum crop biodiversity conservation followed by the respondent is 0.333 while the minimum is 0.238. The mean of crop biodiversity conservation is 0.271 with a standard deviation of 0.026. The coefficient of variance is 0.095 percent which indicates that the distribution of the variable is quite consistent.

The distribution pattern of variable, **Climate resilient technology (Y6),** shows that the maximum climate resilient technology followed by the respondent is 6 while the minimum is 1. The mean of climate resilient technology followed is 3.7 with a standard deviation of 1.501. The coefficient of variance is 0.406 percent which indicates that the distribution of the variable is quite consistent.

The distribution pattern of variable, **Ecosystem output (Y7),** shows that the energy output is 66345while the minimum is 33510. The mean of energy output is 46250.85 with a standard deviation of 46250.85. The coefficient of variance is 0.156 percent which indicates that the distribution of the variable is quite consistent.

COEFFICIENT OF CORRELATION: VALUES FOR POOLED SAMPLE

Table 6.B.4 Coefficient of Correlation: Application of organic manure (Y1) and 22 independent variables

Independent Variables	r value
Age (X1)	**0.305***
Education (X2)	**0.484****
Functional Education (X3)	0.110
Size of family members (X4)	0.049
Size of holding (X5)	0.126
Number of fragment (X6)	**0.589****
Average size of fragment(X7)	**0.560****
Mean distance between two fragments(X8)	**0.418****
Size of homestead land(X9)	**0.833****
Yield of Crop(X10)	**0.588****
Marketed surplus(X11)	**0.500****
Land under irrigation (X12)	0.085
Rainfed area (X13)	0.113
Cropping intensity(X14)	0.143
Total amount of seed used (X15)	0.116
Total hours of irrigation given (X16)	0.152
Total amount of chemical fertilizer applied (X17)	0.079
Count of livestock (X18)	**0.521****
On farm Income (X19)	0.001
Communication Variable(X20)	**0.341***
Consumption of coal, fuel wood, fire wood etc.(X21)	**0.813****
Production of cowdung/poultry liter /goat/biogas slurry(X22)	**0.418****

***significant at 0.05 level**

****significant at 0.01 level**

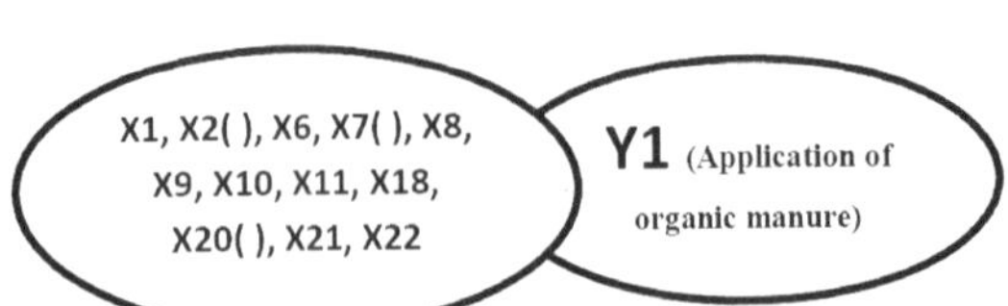

Fig 6.B.4 : Coefficient of correlation: Application of organic manure (Y1) and independent variables having significant correlation.

Fig 1 Presents the coefficient of correlation between Application of organic manure (Y1) and 22 independent variables. It has been found that the following variables Age (X1), Number of fragment (X6), Mean distance between two fragments(X8), Size of homestead land(X9), Yield of Crop(X10), Marketed surplus(X11), Count of livestock (X18), Communication Variable(X20), Consumption of coal, fuel wood, fire wood etc.(X21), Production of cowdung/,poultry liter / goat/biogas slurry(X22) have been recorded significant and positive correlation with the dependent variable, while the other variable Education (X2), Average size of fragment(X7) have recorded significant but negative correlation with the dependent variable.

Revelation

It is discernible that the number of fragments (X6) and the mean distance between two fragments (X8), homestead land (X9), have gone positively and the application of organic manures (). When a size of holding undergoes a process of fragmentation it triggers another process moving in a complex manner with a resultant ramification of input used and its prodigal application. It also helps increasing cost of management as well as decline in input use efficiency thereafter. Count of livestock and production of cow dung, poultry litter and biogas slurry have rightly contributed to the application of higher organic manures. The stewardship function here has got two basic things to do 1) can there be further integration of farm management function and 2) whether this integrations can be ecologically effective efficient when it dovetails higher use of organic manures.

Average size of fragments, as the result suggests, the smaller the size of fragments, the bigger will be the volumes of land resources and logically the higher will be application of organic manure.

The policy implication for this analysis can wait for further results to come , one direction is more apparent to suggest that fragmentation of holding however have boosted the higher application of organic when associated with consumption of fire woods and count of livestock.

Table 6.B.5 Coefficient of correlation: Rice stubble height left in the Field (Y2) and 22 independent variables

Independent Variables	r value
Age (X1)	0.096
Education (X2)	0.008
Functional Education (X3)	0.041
Size of family members (X4)	**0.328***
Size of holding (X5)	0.195
Number of fragment (X6)	0.002
Average size of fragment(X7)	0.047

Independent Variables	r value
Mean distance between two fragments(X8)	0.129
Size of homestead land(X9)	0.107
Yield of Crop(X10)	**0.366****
Marketed surplus(X11)	**0.463****
Land under irrigation (X12)	0.086
Rainfed area (X13)	0.185
Cropping intensity(X14)	0.226
Total amount of seed used (X15)	0.166
Total hours of irrigation given (X16)	0.240
Total amount of chemical fertilizer applied (X17)	**0.361****
Count of livestock (X18)	0.050
On farm Income (X19)	0.252
Communication Variable(X20)	0.001
Consumption of coal, fuel wood, fire wood etc.(X21)	0.033
Production of cowdung/poultry liter /goat/biogas slurry(X22)	0.255

***significant at 0.05 level**

****significant at 0.01 level**

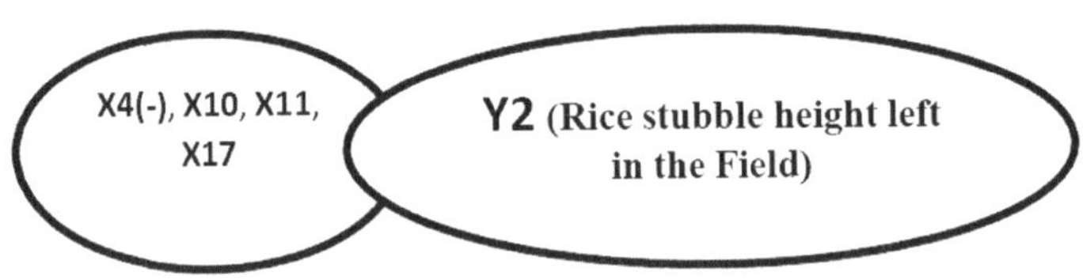

Fig 6.B.5 Coefficient of correlation: Rice stubble height left in the Field (Y2) and independent variables having significant correlation.

Fig 2 Presents the coefficient of correlation between Rice stubble height left in the Field (Y2) and 22 independent variables. It has been found that the following variables Yield of Crop(X10), Marketed surplus(X11), Total amount of chemical fertilizer applied (X17), have been recorded significant and positive correlation with the dependent variable, while the other variable Size of family members (X4), have recorded significant but negative correlation with the dependent variable.

Revelation

When yield comes up better than some stubbles as left over crop residues are kept with the operating land that is how other economic factor like marketed surplus has also been positively associated with length and volume of stubbles left in the field. When agro chemicals are applied at higher level the general perception goes that its effect should be countervailed with the application of organic manures and this is a kind of inverse development , as in cases where perceived negatively are outplayed by perceived positively as well.

The higher size of family has got a systemic need for higher volume of biofuels in rural India and has got a negative impact on stubble height. The burning of stubble is a common practice for parboiling of rice and again it is well visible for the farm families having large number of family member and acute crisis for fuels to do the basic and essential domestic as well as economic function.

Table 6.B.6 Coefficient of correlation: Crop residue left in the field (Y3) and 22 independent variables

Independent Variable	r value
Age (X1)	0.190
Education (X2)	**0.295***
Functional Education (X3)	0.073
Size of family members (X4)	0.043
Size of holding (X5)	0.115
Number of fragment (X6)	**0.384****
Average size of fragment(X7)	0.183
Mean distance between two fragments(X8)	0.266
Size of homestead land(X9)	**0.394****
Yield of Crop(X10)	0.194
Marketed surplus(X11)	0.166
Land under irrigation (X12)	**0.327***
Rainfed area (X13)	0.032
Cropping intensity(X14)	0.162
Total amount of seed used (X15)	0.067
Total hours of irrigation given (X16)	0.125
Total amount of chemical fertilizer applied (X17)	0.024
Count of livestock (X18)	0.189
On farm Income (X19)	0.028
Communication Variable(X20)	0.159
Consumption of coal, fuel wood, fire wood etc.(X21)	**0.331***
Production of cowdung/poultry liter /goat/biogas slurry(X22)	0.162

***significant at 0.05 level**

****significant at 0.01 level**

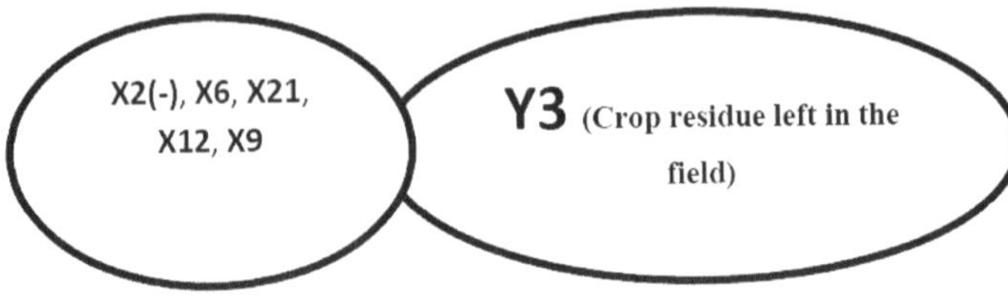

Fig 6.B.6 Coefficient of correlation: Crop residue left in the field (Y3) and independent variables having significant correlation.

Fig 3 Presents the coefficient of correlation between Crop residue left in the field (Y3) and 22 independent variables. It has been found that the following variables Number of fragment (X6), Size of homestead land (X9), Land under irrigation (X12), Consumption of coal, fuel wood, fire wood etc. (X21) have been recorded significant and positive correlation with the dependent variable, while the other variable Education (X2) have recorded significant but negative correlation with the dependent variable.

Revelation

It is well discernible that the more is the number of fragments, the higher has been the crop residues added to the crop land. With higher count of fragment, the farmers have been interested and atuned to high intensity and high diversity farm enterprises; the highest contribution are mostly being made by vegetables and pulses so also has been for where size of homestead land and land under irrigation. The whole lot of holding ecology has been an incomplete orchestration with integrating farm enterprises to support the stewardship function with active participation of family members, especially with farm women having concomitant action both for farm and family chores like sibling care etc.

Consumption of coal, fuel wood, fire wood etc can well be relegated to the green source energy which is a good estimator for having dents forwards stewardship.

So, formal education here has a little to do with stewardship function rather traditional community knowledge and a bit of ecological consciousness have been much effective than the formal schooling or formal institutional informative education.

Table 6.B.7 Coefficient of correlation: Soil conservation level (Y4) and 22 independent variables

Independent Variable	r value
Age (X1)	0.010
Education (X2)	0.190
Functional Education (X3)	0.026
Size of family members (X4)	**0.311***

Independent Variable	r value
Size of holding (X5)	0.237
Number of fragment (X6)	0.226
Average size of fragment(X7)	0.144
Mean distance between two fragments(X8)	0.253
Size of homestead land(X9)	0.101
Yield of Crop(X10)	**0.398****
Marketed surplus(X11)	**0.487****
Land under irrigation (X12)	**0.530**
Rainfed area (X13)	0.255
Cropping intensity(X14)	0.181
Total amount of seed used (X15)	**0.309**
Total hours of irrigation given (X16)	0.266
Total amount of chemical fertilizer applied (X17)	**0.487****
Count of livestock (X18)	0.031
On farm Income (X19)	0.241
Communication Variable(X20)	0.073
Consumption of coal, fuel wood, fire wood etc.(X21)	0.006
Production of cowdung/poultry liter /goat/biogas slurry(X22)	**0.333***

***significant at 0.05 level**

****significant at 0.01 level**

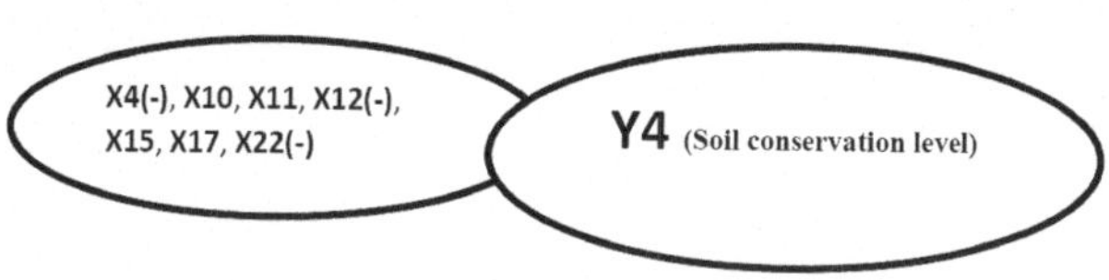

Fig 6.B.7 Coefficient of correlation: Soil conservation level (Y4) and independent variables having significant correlation.

Fig 4 Presents the coefficient of correlation between Soil conservation level (Y4) and 22 independent variables. It has been found that the following variables,Yield of Crop(X10), Marketed surplus(X11), Total amount of seed used (X15), Total amount of chemical fertilizer applied (X17) have been recorded significant and positive correlation with the dependent variable, while the other variable Size of family members (X4), Land under irrigation (X12), Production of cowdung/,poultry liter /goat/biogas slurry(X22) have recorded significant but negative correlation with the dependent variable.

Revelation

The soil conservation interventions, mostly traditional and based on community learning made by the farmers, have been highly effective in maintaining soil health and intrinsic production functions. This community learning, as being perceived, has been refined further with modern soil conservation techniques as well. These include mulching, land shaping, seepage tanks or conservation tanks etc.

So, it's discernible that having this ecological intervention in place, the yield of crop vis a vis marketed surplus, seeds applied and chemical input used, could have performed productively.

This kind of stewardship functions have been effective well for the farm family with small size surrounded by irrigated small sized agro ecosystem and livestock enterprises as well.

Table 6.B.8 Coefficient of correlation: Crop Biodiversity conservation (Y5) and 22 independent variables

Independent variable	r value
Age (X1)	0.231
Education (X2)	0.271
Functional Education (X3)	0.091
Size of family members (X4)	**0.292***
Size of holding (X5)	0.061
Number of fragment (X6)	0.270
Average size of fragment(X7)	**0.308***
Mean distance between two fragments(X8)	0.128
Size of homestead land(X9)	**0.428****
Yield of Crop(X10)	**0.428****
Marketed surplus(X11)	**0.458****
Land under irrigation (X12)	0.133
Rainfed area (X13)	0.057
Cropping intensity(X14)	0.153
Total amount of seed used (X15)	0.200
Total hours of irrigation given (X16)	0.064
Total amount of chemical fertilizer applied (X17)	0.228
Count of livestock (X18)	**0.339***
On farm Income (X19)	0.225
Communication Variable(X20)	0.151
Consumption of coal, fuel wood, fire wood etc.(X21)	**0.307***
Production of cowdung/poultry liter /goat/biogas slurry(X22)	0.144

*significant at 0.05 level

**significant at 0.01 level

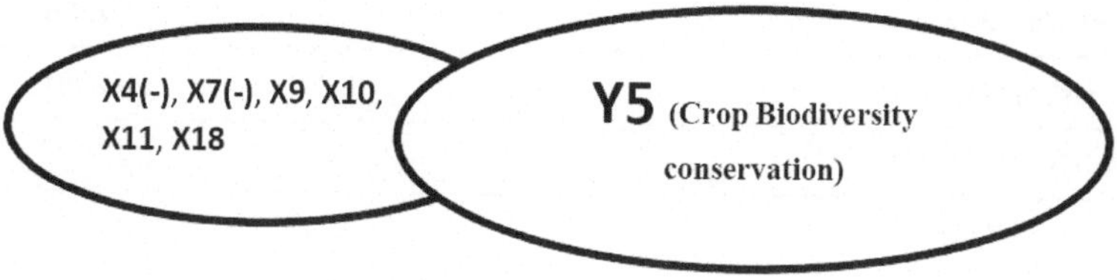

Fig 6.B.8 Coefficient of correlation: Crop Biodiversity conservation (Y5) and independent variables having significant correlation.

Fig 5 Presents the coefficient of correlation between Crop Biodiversity conservation (Y5) and 22 independent variables. It has been found that the following variables Size of homestead land(X9), Yield of Crop(X10), Marketed surplus(X11), Count of livestock (X18) have been recorded significant and positive correlation with the dependent variable, while the other variable Size of family members (X4), Average size of fragment(X7), have recorded significant but negative correlation with the dependent variable.

Revelation

It is nicely delineating that the bigger size homestead land has been found suitable for sustaining higher biodiversity and better crop yield as well. Therefore, marketed surplus has moved upto an upper echelon. The higher crop biodiversity and livestock count have been found mutually synchronized for the villagers under study. To sustain livestock we need to incorporate fodder and wild grasses into the existing cropping sequence. Hence, the crop biodiversity, as a whole, will go up further.

The rise of family members and average size of fragment here have been found not to impact on the crop biodiversity conservation.

Table 6.B.9 Coefficient of correlation: Climate resilient technology (Y6) and 22 independent variables

Independent Variable	r value
Age (X1)	0.189
Education (X2)	0.034
Functional Education (X3)	0.175
Size of family members (X4)	0.069
Size of holding (X5)	0.081
Number of fragment (X6)	**0.282***
Average size of fragment(X7)	0.194
Mean distance between two fragments(X8)	0.118

Independent Variable	r value
Size of homestead land(X9)	**0.296***
Yield of Crop(X10)	0.214
Marketed surplus(X11)	0.206
Land under irrigation (X12)	0.117
Rainfed area (X13)	0.027
Cropping intensity(X14)	0.097
Total amount of seed used (X15)	0.086
Total hours of irrigation given (X16)	0.002
Total amount of chemical fertilizer applied (X17)	0.020
Count of livestock (X18)	0.187
On farm Income (X19)	0.025
Communication Variable(X20)	0.085
Consumption of coal, fuel wood, fire wood etc.(X21)	**0.301***
Production of cowdung/poultry liter /goat/biogas slurry(X22)	0.104

***significant at 0.05 level**

****significant at 0.01 level**

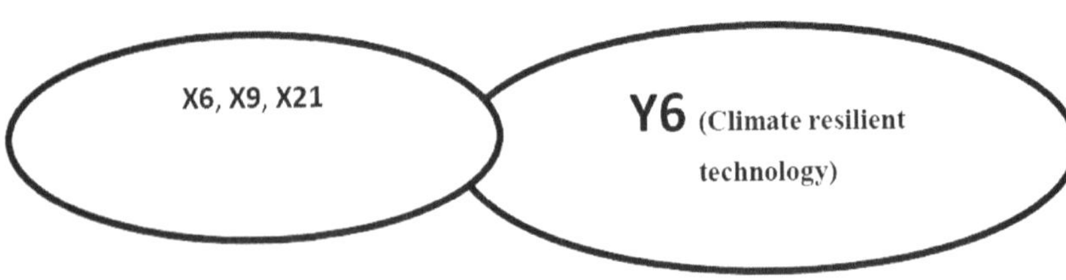

Fig 6.B.9 Coefficient of correlation: Climate resilient technology (Y6) and independent variables having significant correlation.

Fig 6 Presents the coefficient of correlation between Climate resilient technology (Y6) and 22 independent variables. It has been found that the following variables Number of fragment (X6), Size of homestead land (X9), Consumption of coal, fuel wood, fire wood etc. (X21)have been recorded significant and positive correlation with the dependent variable.

Revelation

The farmers for this area are gradually coming up with stewardship function by integrating family practices with the changing scenario of global warming and climate change behavior; may be both under compulsion out of consciousness, they are conserving water, soil as well as biota to make the ecological deliverables there from effective , productive and ecologically sustainable. The ecological dividence, has been reflected through increased yield and escalated marketable surplus, have started generate in the form of mere calibrated stewardship function.

The direct impact of fragments, size of domesticated land and consumption of green energy sources have been well understood by the farmers in terms of their ecological function they are catering to making there agro ecosystem more resilient to cope up with the brunt of global warming and climate change.

Table 6.B.10 Coefficient of correlation: Eco system output (Y7) and 22 independent variables

Independent variable	r value
Age (X1)	**0.367****
Education (X2)	**0.426****
Functional Education (X3)	0.128
Size of family members (X4)	0.302
Size of holding (X5)	0.176
Number of fragment (X6)	**0.506****
Average size of fragment(X7)	**0.342***
Mean distance between two fragments(X8)	0.135
Size of homestead land(X9)	**0.767****
Yield of Crop(X10)	**0.991****
Marketed surplus(X11)	**0.928****
Land under irrigation (X12)	0.103
Rainfed area (X13)	0.041
Cropping intensity(X14)	0.240
Total amount of seed used (X15)	**0.420****
Total hours of irrigation given (X16)	0.031
Total amount of chemical fertilizer applied (X17)	**0.389****
Count of livestock (X18)	**0.470****
On farm Income (X19)	0.267
Communication Variable(X20)	0.153
Consumption of coal, fuel wood, fire wood etc.(X21)	**0.596****
Production of cowdung/poultry liter /goat/biogas slurry(X22)	0.027

***significant at 0.05 level**

****significant at 0.01 level**

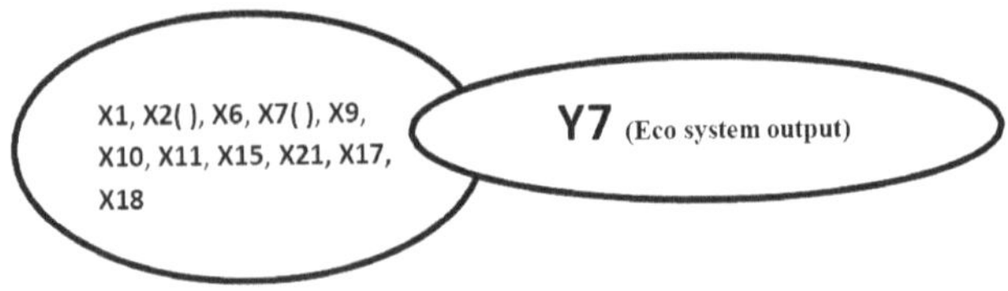

Fig 6.B.10 Coefficient of correlation: Eco system output (Y7) and independent variables having significant correlation.

Fig 7 Presents the coefficient of correlation between Eco system output (Y7) and 22 independent variables. It has been found that the following variables Age (X1), Number of fragment (X6), Size of homestead land(X9), Yield of Crop(X10), Marketed surplus (X11), Total amount of seed used (X15), Total amount of chemical fertilizer applied (X17), Count of livestock (X18), Consumption of coal, fuel wood, fire wood etc. (X21), have been recorded significant and positive correlation with the dependent variable, while the other variable Education (X2), Average size of fragment (X7) have recorded significant but negative correlation with the dependent variable.

Revelation

Ecosystem output has been predicted here out of resource input and consumption variables. The land resources are the basic both for generation and trapping of energy vis a vis energy capsules. Green surroundings with diverse biological and physical entities have well been scripted with the energy behaviours of farmers. The higher stewardship functions stand for efficient energy management and efficient energy management demands and merits effective ecological functions catered through stewardship functions. The same formal education has not been that effective for exerting stewardship function and also smaller size of fragments have been found more effective in delivering stewardship functions than the bigger ones. Simply because, a small consolidated mass appeals intensely to the ecological commitment of the farmers to protect its small but precious land mass from the pernicious effect of erodibility and impoverishment.

Regression Analysis

Table 6.B.11 Step Down Regression Analysis of application of organic manure (Y1) and 22 Causal Variables. The causal variable selected at the last step (step19)

Character	Unstandardized Coefficients		Standardized Coefficients	T value	Sig.
	B	Std. Error	Beta		
X9	384.817	58.948	.589	6.528	.001
X12	2102.113	979.977	.092	2.145	.034
X21	16.976	6.657	.255	2.550	.012
X22	16.881	6.747	.129	2.502	.013

(R = 0.863, R2 = 0.744, Adjusted R2 =0 .736)

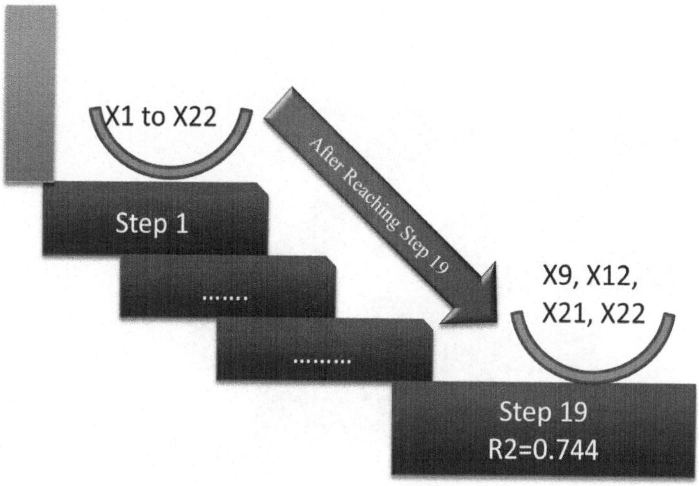

Fig 6.B.11 Step Down Regression Model between Application of organic manure (Y1) and 22 Causal Variables

Results and revelation:

Table 6.B.11presents the step down regression analysis to screen out the critical variables have been, that retained at the last step (19th).

So, Size of homestead land(X9), Land under irrigation (X12), Consumption of coal, fuel wood, fire wood etc.(X21), Production of cowdung/,poultry litter / goat/biogas slurry(X22) are the 4 most important causal variable to interpret the variance embedded with the Application of organic manure (Y1).

Revelation:

So, it reveals that the land and production of bio fuels or its consumption either here functionally critical in ushering the application of organic manure and ultimately goes to scale up stewardship functions for the land under cultivation. This also indicates a need for switching over from chemicalized agriculture to organic farming, even for the owner of the small and fragmented holdings.

Table 6.B.12 Step Down Regression Analysis of Rice stubble height left in the Field (Y2), and 22 Causal Variables. The causal variable selected at the last step (step19)

Character	Unstandardized Coefficients		Standardized Coefficients	T value	Sig.
	B	Std. Error	Beta		
X6	.106	.035	.247	3.109	.002
X11	3.833E 5	.001	.550	7.226	.001
X13	3.197	.929	.241	3.442	.001
X16	.073	.035	.147	2.080	.039

(R = 0.578, R2 = 0.334, Adjusted R2 =0 .316)

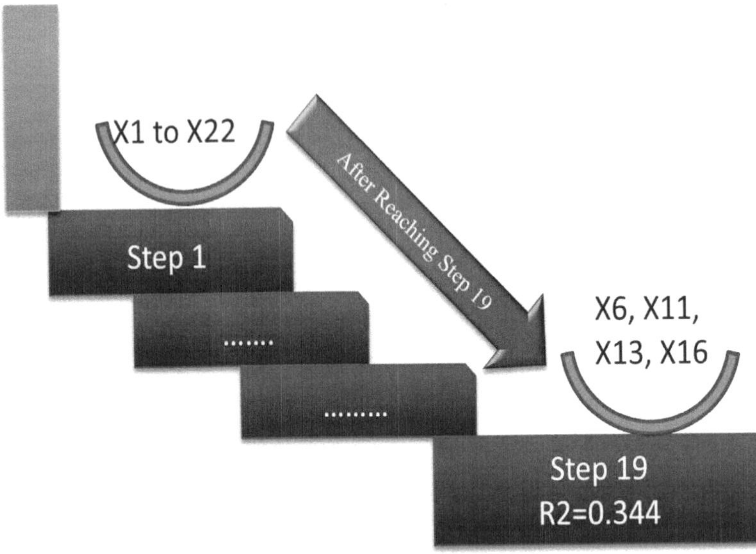

Fig 6.B.12 Step-Down Regression Model Between Rice stubble height left in the Field (Y2) and 22 Causal Variables

Results and revelation:

Table 6.B.12 presents the step down regression analysis to screen out the critical variables that have been retained at the last step (19th).

So, Number of fragment(X6), Marketed surplus(X11), Rainfed area (X13), Total hours of irrigation given (X16), are the 4 most important causal variables to interpret the variance embedded with the Rice stubble height left in the Field (Y2).

Rice stubble height left in the crop field by the farmers is an important indicator to estimate stewardship function both quantitatively and qualitatively. The length of stubble height can keep on adding organic manures, helps higher level of water retention and expedited metabolic function of bacterial scores harbouring inside the soil echelon. The four critical variables, causally impacting on this consequent variable, have been Number of fragment (X6), Marketed surplus(X11), Rainfed area (X13), and Total hours of irrigation given (X16). The number of fragment has well been relegated to managing stubble height supported by marketed surplus and total hours of irrigation given. The amount of water to be drenched in the field in a rain fed agro ecosystem, complex, diverse and risk prone, has got critical functions for its proper decomposition and mixing with the soil mass.

These four causal variables together have explained around 32 per cent of variance embedded with the consequent variable, rice stubble height left in the field.

There may be some hidden factors having a low level flamboyance, could have been elicited to characterize the causal interaction. So, inclusion of some more functioning variables could have added higher level of explicability.

Table 6.B.13 Step Down Regression Analysis of Crop residue left in the field (Y3), and 22 Causal Variables. The causal variable selected at the last step (step 20)

Character	Unstandardized Coefficients		Standardized Coefficients	T value	Sig.
	B	Std. Error	Beta		
X2	112.434	51.519	.189	2.182	.031
X9	74.716	27.150	.241	2.757	.007
X12	3224.656	840.379	.298	4.009	.001

(R = 0.498, R2 = 0.248, Adjusted R2 =0.233)

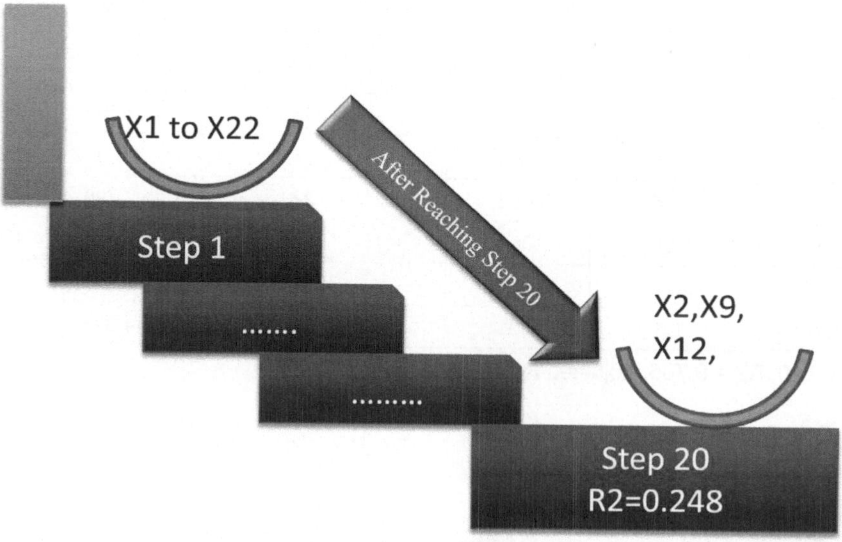

Fig 6.B.13 Step Down Regression Model Between Crop residue left in the field (Y3) and 22 Causal Variables

Results and revelation:

Table 6.B.13 presents the step down regression analysis to elicit the critical variables, out of the whole plethora of variables which have been retained at the last step (20th) to contribute substantially on the consequent variable, that is, Crop residue left in the field (Y3).

So, Education (X2), Size of homestead land(X9), Land under irrigation (X12) are the three most important causal variables to interpret the variance embedded with the Crop residue left in the field (Y3).

Crop residues are important by products about which most of the farmers remain either stoic or indifferent. This indicates a low level of stewardship function which could have been otherwise a great contribution for conserving soil health and water. It has been found that three variables Education (X2), Size of homestead land(X9), Land under irrigation (X12) have been retained at the last step but have explained as much as 24.8 per cent of variance. Education adds to ecological sense and stewardship responsibility, which here has been supported by land under irrigation and size of homestead land.

For this study area homestead land are successfully utilized for crop enterprise like vegetables, orchard and plantations which, again, is the source of huge green residues for maintaining the soil health and ecological functions therein.

Table – 6.B.14 Step Down Regression Analysis of Soil conservation level (Y4), and 22 Causal Variables. The causal variable selected at the last step (step 16)

Character	Unstandardized Coefficients		Standardized Coefficients	T value	Sig.
	B	Std. Error	Beta		
X1	.001	.001	.099	1.959	.052
X4	.028	.005	.401	6.276	.001
X5	1.663	.149	.687	111.133	.001
X9	.005	.002	.223	2.796	.006
X11	3.760E 6	.001	.542	5.501	.001
X14	.001	.001	.146	2.883	.005
X17	.001	.001	.309	4.353	.001

(R = 0.840, R2 = 0.705, Adjusted R2 =0.690)

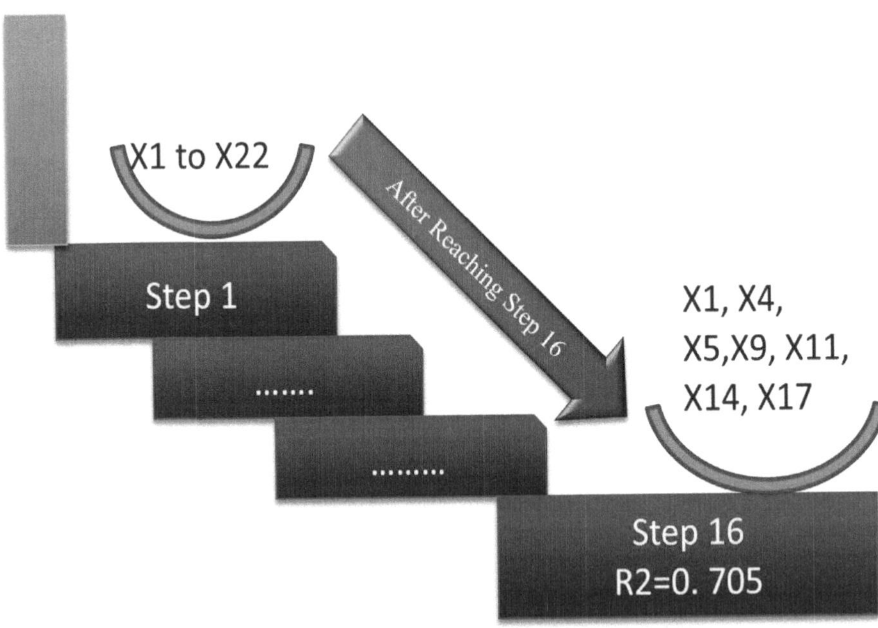

Fig 6.B.14 Step Down Regression Model Between Soil conservation level (Y4) and 22 Causal Variables Results and revelation:

Table 6.B.14 presents the step down regression analysis presents the critical Variables out of the whole plethora of variables, have been retained at the last step (16th) to contribute substantially on the consequent variable that is Soil conservation level (Y4).

So, Age (X1), Size of family members (X4), Size of holding (X5), Size of homestead land(X9), Marketed surplus(X11),Cropping intensity(X14) , Total amount of chemical fertilizer applied (X17), are the 7 most important causal variable to interpret the variance embedded with the Soil conservation level (Y4). Yiridoe et. al in 2010 in regression analysis found that farm characteristics (i.e., farm type, farm size, farm income) and farmer capacity variables (i.e., specialized training and knowledge from EFP (Nova Scotia's Environmental Farm Plan) program information sessions and workshops, and on farm stewardship demonstrations) were significant determinants of environmental farm planning.

Soil conservation level, being a critical stewardship function, has to go beyond conservation itself. Crop biodiversity, soil moisture retention, invisible microbial metabolic function, yield behavior of crop and quality, land shaping and landscaping are all important deliverables out of soil conservation, keep on directing and maneuvering ecological functions to support both human and non human life forms.

The following variables Age (X1), Size of family members (X4), Size of holding (X5), Size of homestead land(X9), Marketed surplus(X11),Cropping intensity(X14) , Total amount of chemical fertilizer applied(X17) have been retained to contribute as high as 69% of variance embedded the dependent variable Soil conservation level (Y4). These causal variables can be clubbed into three categories that are 1) Socio economic by nature, 2) Biodiversity index, 3) Input management in nature. When combination of causal variables encompasses three vital aspects of stewardship function no wonder the variance explained will go up substantially.

Table 6.B.15 Step Down Regression Analysis of Crop Biodiversity conservation (Y5), and 22 Causal Variables. The causal variable selected at the last step (step 17)

Character	Unstandardized Coefficients		Standardized Coefficients	T value	Sig.
	B	Std. Error	Beta		
X4	.004	.001	.291	3.122	.002
X5	.151	.049	.301	3.094	.002
X9	.002	.001	.379	4.480	.001
X13	.053	.021	.194	2.570	.011
X18	.005	.002	.246	2.454	.015
X22	.001	.001	.312	3.818	.001

($R = 0.607$, $R2 = 0.368$, Adjusted $R2 = 0.341$)

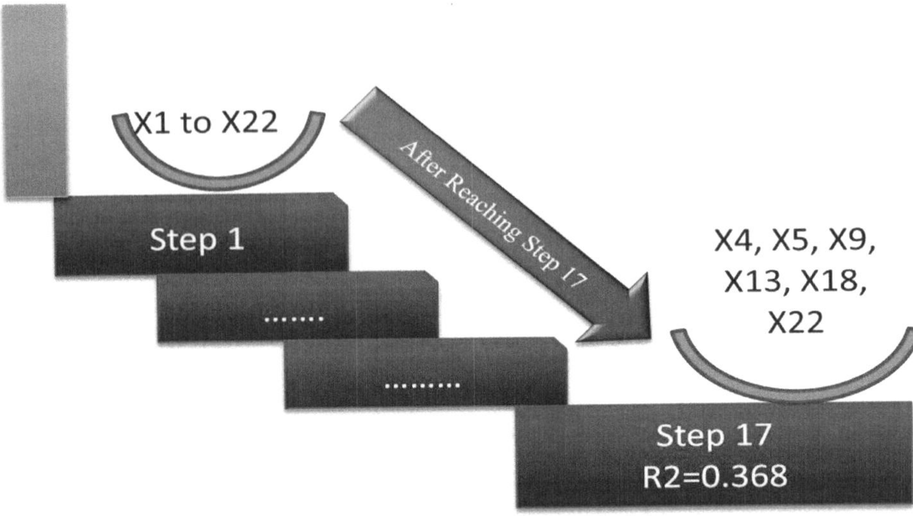

Fig 6.B.15 Step Down Regression Model Between Crop Biodiversity conservation (Y5) and 22 Causal Variables

Results and revelation

Table 6.B.15 presents the step down regression analysis to screen out the critical variables that have been retained at the last step (17th).

So, Size of family members (X4), Size of holding (X5), Size of homestead land(X9), Rainfed area (X13), Count of livestock (X18), Production of cowdung/,poultry litter /goat/biogas slurry(X22).are the 6 most important causal variable to interpret the variance embedded with the Crop Biodiversity conservation (Y5).

Crop biodiversity has been the building block to configure the ecological health to such a functional level and, driving and directing the productive function of the operating agro ecosystem to a desired level. Several studies showed that the species diversity in homestead garden can range from less than 5 to more than 100. When diversity calculated through Simpson Index for herb species, the result was similar for rural and urban homestead garden but for shrubs. It was urban homesteads which scored higher in a study at Bangladesh. Here homestead garden has played the most effective and widespread measure of biodiversity as the count of natural forest is decreasing day by day. (Biswajit Roy et.al. 2013)

When the member of a farm family is large by count and culture, habit and choices, the crop biodiversity moves up flamboyantly to add to a count of ecological functions and get them driven to serve anthropogenic need. The resource variables, family members (X4), Size of homestead land(X9), Rainfed area (X13), Count of livestock (X18), Production of cowdung/poultry litter/goat/biogas slurry(X22) all have got a competing and composite impact on one of the most ecological function that is Crop Biodiversity conservation (Y5).

To nurture and cherish more and different crop per unit area per cropping season a farmer has to have options and choices for different types of land locating in different topographic expanse. So, also has been true for livestock and land under rain fed agriculture as well.

Table—6.B.16 Step Down Regression Analysis of Climate resilient technology (Y6) and 22 Causal Variables. The causal variable selected at the last step (step 19)

Character	Unstandardized Coefficients		Standardized Coefficients	T value	Sig.
	B	Std. Error	Beta		
X7	4.563	2.066	.186	2.208	.029
X11	1.685E 5	.001	.272	2.924	.004
X12	1.292	.607	.173	2.127	.035
X15	0.019	.007	.243	2.716	.007

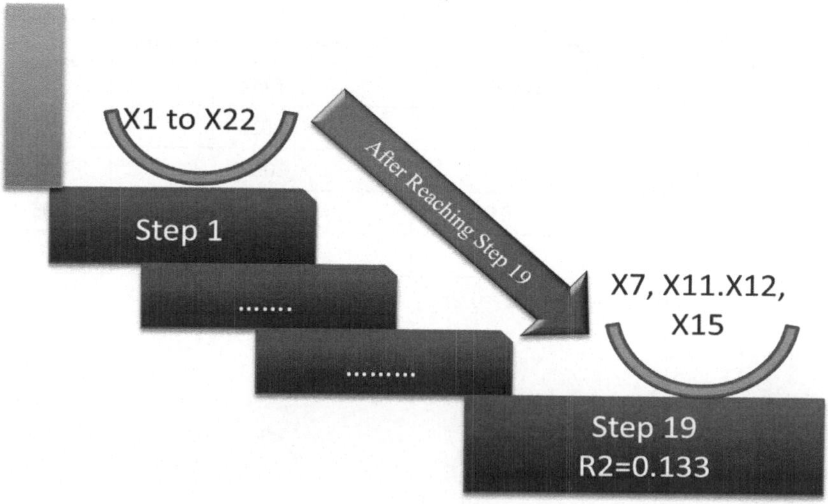

(R = 0.365, R2 = 0.133, Adjusted R2 =0.109)

Fig 6.B.16 Step Down Regression Model Between Climate resilient technology (Y6) and 22 Causal Variables

Results and revelation:

Table 6.B.16 presents the step down regression analysis to screen out the critical variables that have been retained at the last step (19th).

So, Average size of fragment(X7), Marketed surplus(X11), Land under irrigation (X12), Total amount of seed used (X15), are the 4 most important causal variable to interpret the variance embedded with the Climate resilient technology (Y6).

Climate resilient technology imparts and implicates the kind and score of technologies which has got ability to absorb climatic shock and cope up climate variability. This type of technologies is having both integrating and segregating function. So, that an optimum productivity can be obtained, nevertheless there is harsh and sultry weather in and around. When extreme events in a system are tamed by elasticity of input and technological characters, the resilience of the system goes up to support expected production function, may be branded as an optimal function theory.

The four variables Average size of fragment(X7), Marketed surplus(X11), Land under irrigation (X12), Total amount of seed used (X15), have together contributed just 13.3% of the variance embedded resilience character of the technology.

Since research and testing on climate resilient technologies are at the nascent stage, both the perception and experience about is very low amongst the farmers. Hence, the variance explained runs at low level.

Table—6.B.17 Step Down Regression Analysis of Ecosystem output (Y7) and 22 Causal Variables. The causal variable selected at the last step (step 19)

Character	Unstandardized Coefficients		Standardized Coefficients	T value	Sig.
	B	Std. Error	Beta		
X2	111.807	51.235	.026	2.182	.031
X9	115.209	36.766	.051	3.134	.002
X10	.975	.015	.940	63.996	.001
X12	3086.405	804.444	.039	3.837	.001

(R = 0.993, R2 = 0.986, Adjusted R2 =0.986)

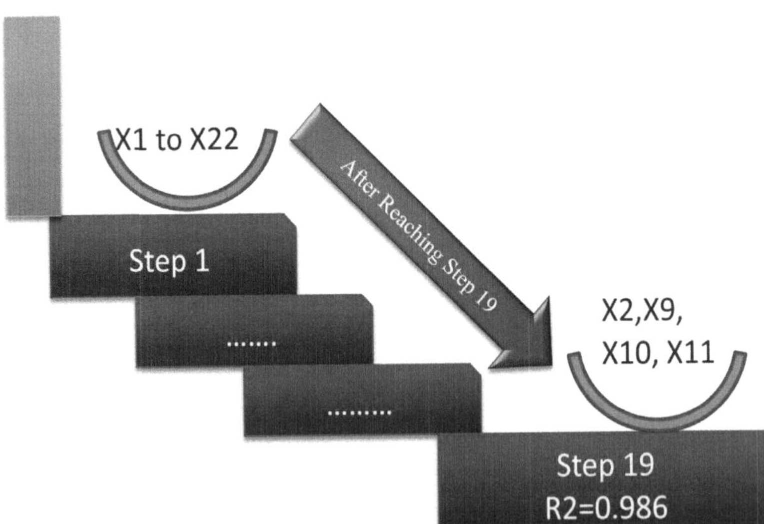

Fig 6.B.17 Step Down Regression Model Between Energy output (Y7) and 22 Causal Variables

Results and revelation:

Table 6.B.17 presents the step down regression analysis to screen out the critical variables that have been retained at the last step (19th).

So, Education (X2), Size of homestead land(X9), Yield of Crop(X10), Land under irrigation (X12), are the 4 most important causal variable to interpret the variance embedded with the Ecosystem output (Y7).Ecosystem output here has been measured in terms of green and raw yield from the field. When soil is managed properly, water is retained adequately across the months and management is efficient, output comes up amply and effectively. Here the variables Education (X2), Size of homestead land(X9), Yield of Crop(X10), Land under irrigation (X12) conspicuous by land resources , marketed surplus and applied seed input have contributed substantially to usher both the quantity and quality of system output. To have desired system output we need integration and orchestration of factors and functions prevailing and performing in any ecosystem. Here the biological output has well been explained by the contribution of these four resources and management factors and has contributed swash buckling 98.6% variance.

6.B.18 Dendogram analysis:

G1 7,46,15,33,11,31,27,32,38,24,42,45,40,23,49,35,124,9,18,37,4,3,17,10,39,26,5,482 0,36,13,103

G2-12,34,25,123,130,8,22,21,1,16,146,104,145,44,50,43,105,114

G3 28,129,142,131,47,108,113,6,41,106,125,14,144,19,30,2,29,143,141,126,112,132,1 34,110.

G4- 115, 117, 109, 150,107,136,147,102,116, 135, 139, 137, 148, 140, 133, 149

G5-119, 127, 111, 120, 118, 121, 138, 122, 101

G6-52, 54, 67

Revelation

Dendogram is a strategic clustering of respondents based on their homogeneity and proximity may be within village and across the village. This is a graphical presentation of cluster analysis wherein groups are created based on Euclidean distances, that is the higher is the distance between groups, the higher would be the difference between/ amongst the groups. Here, we can trace out six homogeneous groups which would shapeup grouping of respondents for different kind of participatory interventions and capacity building e.g; Training, Planning, Performance appraisal, Creation of entrepreneurial groups, Natural resource management task force, Stewardship functions and peer groups.

Table 6.B.19Factor analysis: Village 1 (Panchkahaniya)

Factor		Factor loading	Variance (%)	Cumulative Variance	Factor Renaming
Factor 1	Education (X2)	0.399	19.551	19.551	**Resource Status**
	Size of holding (X5)	0.953			
	Number of fragment (X6)	0.759			
	Land under irrigation (X12)	0.837			
	Rainfed area (X13)	0.784			
Factor 2	Size of family members (X4)	0.653	11.878	31.429	**Family Capacity**
	Total hours of irrigation given (X16)	0.264			
	On farm Income (X19)	0.451			
Factor 3	Mean distance between two fragments(X8)	0.441	9.709	41.138	**Family Productive Function**
	Cropping intensity(X14)	0.435			
	Count of livestock (X18)	0.547			
	Communication Variable(X20)	0.417			
	Consumption of coal, fuel wood, fire wood etc.(X21)	0.533			
	Production of cowdung/ poultry liter /goat/biogas slurry(X22)	0.754			
Factor 4	Age (X1)	0.524	9.123	50.260	**Farm entrepreneurship**
	Size of homestead land(X9)	0.430			
	Yield of Crop(X10)	0.434			
	Marketed surplus(X11)	0.458			

Table 6.B.19 presents the factor analysis for the conglomeration of homogeneous variables based on factor loading and Eigen values into a factor, which has been renamed thereafter. Factor 1 has accommodated the following variables, Education (X2), Size of holding (X5), Number of fragment (X6), and Land under irrigation (X12), Rainfed area (X13). These factors have contributed 19.55 per cent variance and have been named as Resource Status. Factor 2 has accommodated the following variables Size of family members (X4), Total hours of irrigation given (X16), On farm Income (X19). These factors have contributed 11.88 per cent variance and have been

renamed as Family Capacity. Factor 3 has accommodated the following variables Mean distance between two fragments (X8), Cropping intensity (X14), Count of livestock (X18), Communication Variable (X20), Consumption of coal, fuel wood, fire wood etc. (X21), Production of cowdung/ poultry liter /goat/biogas slurry (X22). These factors have contributed 9.70 per cent variance and have been renamed as Family Productive Function. Factor 4 has accommodated the following variables Age (X1), Size of homestead land(X9), Yield of Crop (X10), Marketed surplus (X11). These variables have contributed 9.12 per cent of variance and have been renamed as Farm entrepreneurship.

Table 6.B.20 Factor Analysis: Village 2 (Bhabanipur)

Factor VILLAGE 2		Factor loading	Variance (%)	Cumulative Variance	Factor Renaming
Factor 1	Education (X2)	0.402	15.059	15.059	**Farm Modernity**
	Size of holding (X5)	0.817			
	Average size of fragment(X7)	0.536			
	Land under irrigation (X12)	0.775			
	Total amount of chemical fertilizer applied (X17)	0.536			
	Count of livestock (X18)	0.661			
Factor 2	Yield of Crop(X10)	0.768	13.296	28.355	**Crop Entrepreneurship**
	Marketed surplus(X11)	0.718			
Factor 3	Rainfed area (X13)	0.441	10.748	39.103	**Energy Resource Dyad**
	Consumption of coal, fuel wood, fire wood etc.(X21)	0.5452			
Factor 4	Functional Education (X3)	0.308	8.516	47.619	**Entrepreneurial Communication**
	Communication Variable(X20)	0.449			
Factor 5	Number of fragment (X6)	0.472	8.015	55.634	**Crop Economy**
	Total amount of seed used (X15)	0.483			
	On farm Income (X19)	0.573			
Factor 6	Size of family members (X4)	0.601	6.725	62.359	**Family Resource**
	Size of homestead land(X9)	0.528			
Factor 7	Mean distance between two fragments(X8)	0.305	6.502	68.861	**Geo social determinants**
	Total hours of irrigation given (X16)	0.371			
	Production of cowdung/ poultry liter /goat/biogas slurry (X22)	0.534			

Table 6.B.20 presents the factor analysis for the conglomeration of homogeneous variables based on factor loading into a factor, which has been renamed thereafter. Factor 1 has accommodated the following variables Education (X2), Size of holding (X5), Average size of fragment(X7), Land under irrigation (X12), Total amount of chemical fertilizer applied (X17), Count of livestock (X18). These factors have contributed 15.059 per cent variance and have been named as Farm Modernity. Factor 2 has accommodated the following variables Yield of Crop(X10), Marketed surplus(X11).

These factors have contributed 13.296 per cent variance and have been renamed as Crop Entrepreneurship. Factor 3 has accommodated the following variables Rainfed area (X13), Consumption of coal, fuel wood, fire wood etc.(X21). These factors have contributed 10.748 per cent variance and have been renamed as Energy Resource Dyad. Factor 4 has accommodated the following variables Functional Education (X3), Communication Variable(X20). These variables have contributed 8.516 per cent of variance and have been renamed as Entrepreneurial Communication. Factor 5 has accommodated the following variables: Number of fragment (X6), On farm Income (X19), Total amount of seed used (X15).These variables have contributed 8.015per cent of variance and have been renamed as Crop Economy. Factor 6 has accommodated the following variables: Size of family members (X4), Size of homestead land(X9).These variables have contributed 6.725 per cent of variance and have been renamed as Family Resource. Factor 7 has accommodated the following variables: Mean distance between two fragments(X8), Total hours of irrigation given (X16), Production of cowdung/ poultry liter /goat/ biogas slurry(X22). These variables have contributed6.502per cent of variance and have been renamed as Geo social determinants.

Table 6.B.21 Factor Analysis : Village 3 (Satyapul)

Factor Village 3		Factor loading	Variance (%)	Cumulative Variance	Factor Renaming
Factor 1	Size of holding (X5)	0.651	16.823	16.823	**Land sociology**
	Average size of fragment(X7)	0.824			
	Rainfed area (X13)	0.257			
	Count of livestock (X18)	0.401			
	On farm Income (X19)	0.566			
Factor 2	Size of family members (X4)	0.576	14.694	31.518	**Family interaction**
	Cropping intensity(X14)	0.505			
	Communication Variable(X20)	0.526			

Factor Village 3		Factor loading	Variance (%)	Cumulative Variance	Factor Renaming
Factor 3	Education (X2)	0.544	9.474	40.992	**Entrepreneurial capacity**
	Yield of Crop(X10)	0.537			
	Marketed surplus(X11)	0.525			
Factor 4	Age (X1)	0.524	9.123	50.260	**Bio social Re- source**
	Number of fragment (X6)	0.430			
	Total amount of seed used (X15)	0.434			
Factor 6	Mean distance between two fragments(X8)	0.457	6.228	63.422	**Spatial Manage- ment**
	Total hours of irrigation given (X16)	0.429			
	Production of cowdung/ poultry liter /goat/biogas slurry(X22)	0.479			
Factor 9	Rainfed area (X13)	0.257	3.947	77.903	**Resource Mo- dernity**
	Total amount of chemical fertilizer applied (X17)	0.550			

Table6.B.21 presents the factor analysis for the conglomeration of homogeneous variables based on factor loading into a factor, which has been renamed thereafter. Factor 1 has accommodated the following variables: Size of holding (X5), Average size of fragment(X7), Rainfed area (X13), Count of livestock (X18) , On farm Income (X19), Total amount of chemical fertilizer applied (X17), Count of livestock (X18). These factors have contributed 16.823 per cent variance and have been named as Land sociology. Factor 2 has accommodated the following variables: Size of family members (X4), Cropping intensity(X14), Communication Variable(X20).These factors have contributed 14.694 per cent variance and have been renamed as Family interaction. Factor 3 has accommodated the following variables: Education (X2), Yield of Crop(X10), and Marketed surplus (X11). These factors have contributed 9.474 per cent variance and have been renamed as Entrepreneurial capacity. Factor 4 has accommodated the following variables Age (X1), Number of fragment (X6), Total amount of seed used (X15). These variables have contributed 9.123 per cent of variance and have been renamed as Bio social Resource. Factor 6 has accommodated the following variables: Mean distance between two fragments(X8), Total hours of irrigation given (X16), Production of cowdung/ poultry liter /goat/biogas slurry(X22).These variables have contributed 6.228 per cent of variance and have been renamed as Spatial Management. Factor 9 has accommodated the following variables: Rainfed area (X13), Total amount of chemical fertilizer applied (X17). These variables have contributed 3.947 per cent of variance and have been renamed as Resource Modernity

Revelation

The villages under study are presenting here three social ecologies having both similarities and uniqueness. Accordingly, the same set of 22 exogenous variables are interacting amongst themselves to go for , based on Euclidean distance and factor loading, homogeneous conglomeration of variables called factor. While for village 1 (Panchkahaniya) factor 1 (19.55 per cent variance contribution) has been renamed as resource status to indicate that the entrepreneurial process for this village has basically been attributed to its resource characters by size of holding or number of fragments. So, land resources here have extended just a resource backup not the entrepreneurial process as such. For village 3 (Satyapul) factor 1 has been renamed as land sociology to justify and elucidate a socialization process already in place to characterize its crop entrepreneurship and its institulization. The same factor 1 for village 2 (Bhabanipur) has been renamed as farm modernity wherein the indication of farm modernization are conspicuous and predominant by their intrusive interaction between land, livestock, irrigation and expectedly upcoming market.

So, three villages can be redeemed and re-envisaged as village with resource domination, another resource with entrepreneurial socialization and resource with modernizing agricultural socialization having entrepreneurial inputs.

For Factor 2, in each of the villages we find three different types of interactions amongst the respective variables. While for village 1 (Panchkahaniya) it is family capacity, the human resources, access to irrigation and income have contributed to family for village 2 (Bhabanipur) it has been crop entrepreneurship and for village 3 (Satyapul) it is family interaction.

So, Family capacity for village has contributed to entrepreneurship and stewardship together, the family interaction in village 2 (Bhabanipur) through a communication network has made the interaction unique as well as productive to go for stewardship as well as economic growth. For village 3 (Satyapul) the stewardship has to deal with crop entrepreneurship , even at the family level three villages have got their own way of family exposition to creative entrepreneurship and ecological deliverables.

For the rest of the factors accordingly we find an ecology related and entrepreneurship driven stewardship behaviours. While family productive functions and farm has been proactive in one village (Satyapul) in terms of stewardship functions; the entrepreneurial capacity, bio social resources and geo spatial management have been relegated to both ecological function and entrepreneurial discourses to define agricultural enterprise in terms of resource stewardship through creating ecological deliverables for appeasing anthropogenic function. Similarly for village Bhabanipur the rest of the factors, crop economy, family resource and entrepreneurial communication have gone deeper into the issues and responsibilities for catering stewardship function.

Path Analysis Table

Table 6.B.22 PATH ANALYSIS: Application of organic manure (Y1) Vs 22 exogenous variables

Variable	Total Effect	Direct Effect	Indirect effect
Size of homestead land (X9)	0.833	.589	.244
Land under irrigation (X12)	0.085	.092	.177
Consumption of coal, fuel wood, fire wood etc. (X21)	0.813	.255	.558
Production of cow dung/ poultry liter /goat/biogas slurry (X22)	0.418	.129	.289
Residual Effect= 25.6 %			

Table 6.B.22 presents the path analysis wherein the total effect has been decomposed into direct, indirect and residual effect. Here the variables retained at the last step of step wise regression analysis have been considered and the respective β values are counted for direct effect exerted by the respective causal variable. The table reveals that the variable Consumption of coal, fuel wood, fire wood etc. (X21) has got higher indirect effect exerted by the rest other variables on it and it is higher than its direct effect. So, energy consumption and its behaviour is mostly characterized by the associational effect of other variables. So, also is happening for another variable Production of cow dung/poultry liter /goat/biogas slurry (X22) where in the contribution of other variables speaks a lot than the variable itself speaks. So far as Application of organic manure (Y1) is concerned.

So stewardship function through Application of organic manure (Y1) has well being predicted by these four variables but the direct effect of at least two variables are less than the indirect effect exerted by other variables on this two.

Table 6.B.23 PATH ANALYSIS: Rice stubble height left in the Field (Y2) Vs 22 exogenous variables

Variable	Total Effect	Direct Effect	Indirect effect
Number of fragment (X6)	0.002	.247	.249
Marketed surplus (X11)	0.463	.550	.087
Rain fed area (X13)	0.185	.241	0.056
Total hours of irrigation given (X16)	0.240	.147	0.093
Residual Effect= 66.6 %			

Table 6.B.23 presents the path analysis wherein the total effect has been decomposed into direct, indirect and residual effect. Here the variables retained at the last step of step wise regression analysis have been considered and the respective β values are counted for direct effect exerted by the respective causal variable. The table reveals that the variable Number of fragment (X6) has got higher indirect effect

exerted by the rest other variables on it and it is higher than its direct effect. So, Count of fragment and its behaviour is mostly characterized by the associational effect of other variables. So far as is Rice stubble height left in the Field (Y2) concerned.

So stewardship function through Rice stubble height left in the Field (Y2) has well being predicted by these four variables but the direct effect of one variable is less than the indirect effect exerted by other variables on this one.

Table 6.B.24 PATH ANALYSIS: Crop residue left in the field (Y3) Vs 22 exogenous variables

Variable	Total Effect	Direct Effect	Indirect effect
Education (X2)	0.295	.189	0.106
Size of homestead land (X9)	0.394	.241	0.153
Land under irrigation (X12)	0.327	.298	0.029
Residual effect =75.2 %			

Table 6.B.24 presents the path analysis wherein the total effect has been decomposed into direct, indirect and residual effect. Here the variables retained at the last step of step wise regression analysis have been considered and the respective β values are counted for direct effect exerted by the respective causal variable. The table reveals that the variable Size of homestead land(X9) has got higher indirect effect exerted by the rest other variables on it .So, homestead area and its behaviour is mostly characterized by the associational effect of other variables. So far asCrop residue left in the field (Y3) is concerned.

So stewardship function through Crop residue left in the field (Y3) has well being predicted by these three variables.

Table 6.B.25 PATH ANALYSIS: Soil conservation level (Y4) Vs 22 exogenous variables

Variable	Total Effect	Direct Effect	Indirect effect
Age (X1)	0.010	.099	0.109
Size of family members (X4)	0.311	.401	.09
Size of holding (X5)	0.237	.687	.45
Size of homestead land (X9) 0.101 .223 0.324			
Marketed surplus (X11),	0.487	.542	0.055
Cropping intensity (X14)	0.181	.146	0.035
Total amount of chemical fertilizer applied (X17)	0.487	.309	0.178
Residual effect=29.5%			

Table 6.B.25 presents the path analysis wherein the total effect has been decomposed into direct, indirect and residual effect. Here, the variables retained at the last step of the step wise regression analysis have been considered and the respective values are counted for direct effect exerted by the respective causal variable. The table reveals that the variable Size of homestead land(X9) has got higher indirect effect exerted by the rest other variables on it and it is higher than its direct effect. So, size of homestead land and its attribution is mostly characterized by the associational effect of other variables. The same also is happening for another two variables Age (X1), Size of holding (X5) where in the contribution of other variables speaks a lot than the variable itself speaks. So far as Soil conservation level (Y4) is concerned.

So stewardship function through Soil conservation level (Y4) has well being predicted by these seven variables but the direct effect of at least four variables are less than the indirect effect exerted by other variables on these four.

Table 6.B.26 PATH ANALYSIS: Crop Biodiversity conservation (Y5) Vs 22 exogenous variables

Variable	Total Effect	Direct Effect	Indirect effect
Size of family members (X4)	0.292	.291	0.001
Size of holding (X5)	0.061	.301	0.362
XSize of homestead land(X9)	0.428	.379	0.049
Rain fed area (X13)	0.057	.194	0.137
Count of livestock (X18),	0.144	.246	0.39
Production of cow dung/poultry liter /goat/biogas slurry(X22).	0.339	.312	0.651
Residual Effect= 63.2 %			

Table 6.B.26 presents the path analysis wherein the total effect has been decomposed into direct, indirect and residual effect. Here, the variables retained at the last step of step wise regression analysis have been considered and the respective values are counted for direct effect exerted by the respective causal variable. The table reveals that the variable Production of cow dung/poultry liter /goat/biogas slurry(X22) has got higher indirect effect exerted by the rest other variables on it and it is higher than its direct effect. So, production of cow dung, poultry liter, biogas slurry and its behaviour is mostly characterized by the associational effect of other variables. So, also is happening for another variable Size of holding (X5) where in the contribution of other variables speaks a lot than the variable itself speaks. So far as Crop Biodiversity conservation (Y5) is concerned.

So stewardship function through Crop Biodiversity conservation (Y5) has well been predicted by these six variables, but the direct effect of at least two variables are less than the indirect effect exerted by other variables on this two.

Table 6.B.27 PATH ANALYSIS: Climate resilient technology (Y6) Vs 22 exogenous variables

Variable	Total Effect	Direct Effect	Indirect effect
Average size of fragment (X7)	0.194	.186	0.008
X11 Marketed surplus (X11)	0.206	.272	0.066
X12 Land under irrigation (X12)	0.117	.173	0.056
Total amount of seed used (X15)	0.086	.243	0.157
Residual Effect = 86.7 %			

Table 6.B.27presents the path analysis wherein the total effect has been decomposed into direct, indirect and residual effect. Here the variables retained at the last step of step wise regression analysis have been considered and the respective values are counted for direct effect exerted by the respective causal variable. The table reveals that the variable Total amount of seed used (X15) has got higher indirect effect exerted by the rest other variables on it and it is higher than its direct effect. So, Total amount of seed used and its behaviour is mostly characterized by the associational effect of other variables. So far as Climate resilient technology (Y6) is concerned.

So stewardship function through Climate resilient technology (Y6) has well being predicted by these four variables but the direct effect of one variable are less than the indirect effect exerted by other variables on this one.

Table 6.B.28 PATH ANALYSIS: Eco system output (Y7) Vs 22 exogenous variables

Variable	Total Effect	Direct Effect	Indirect effect
Education (X2)	0.426	.026	0.4
Size of homestead land (X9)	0.767	.051	0.716
Yield of Crop (X10)	0.991	.940	0.051
Land under irrigation (X12)	0.103	.039	0.064
Residual effect = 1.4 %			

Table 6.B.28 presents the path analysis wherein the total effect has been decomposed into direct, indirect and residual effect. Here the variables retained at the last step of step wise regression analysis have been considered and the respective values are counted for direct effect exerted by the respective causal variable. The table reveals that the variable Size of homestead land (X9) has got higher indirect effect exerted by the rest other variables on it and it is higher than its direct effect. So, Size of homestead land and its behaviour is mostly characterized by the associational effect of other variables. So, also is happening for another variable Land under irrigation (X12) where in the contribution of other variables speak a lot than the variable itself speaks. So far as Eco system output (Y7) is concerned.

So stewardship function through Eco system output (Y7) has well being predicted by these four variables but the direct effect of at least two variables are less than the indirect effect exerted by other variables on this two.

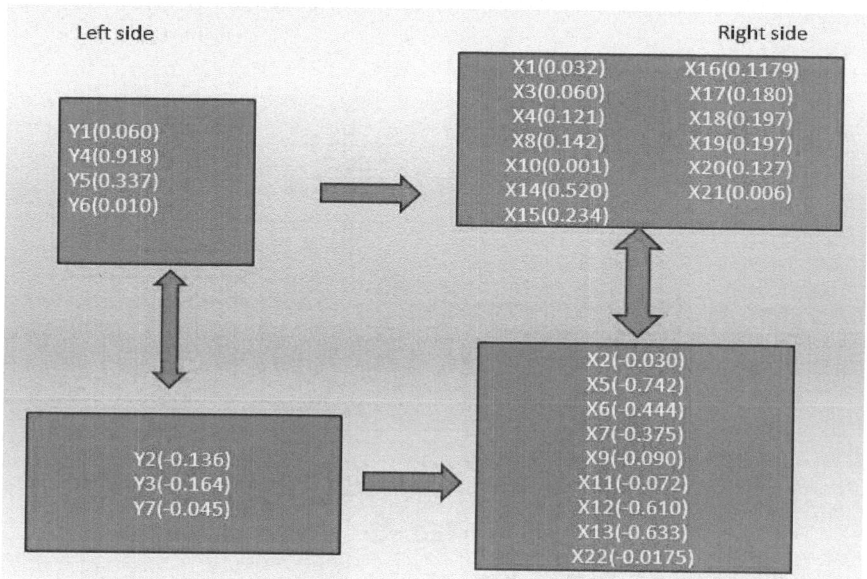

Fig 6.B.29 Canonical Correlation ANALYSIS: VILLAGE 1 (Panchkahaniya)

Results

The paradigm on canonical correlation analysis depicts that the four consequent variables viz, Application of organic manure (Y1),Soil conservation level (Y4),Crop Biodiversity conservation (Y5),Climate resilient technology (Y6)show an affinity to move or interact together in the left side echelon, while they bestow a trend for clandestine selection of right side variable (Age (X1), Functional Education (X3), Size of family members (X4), Mean distance between two fragments(X8), Yield of Crop(X10), Cropping intensity (X14) ,Total amount of seed used (X15), Total hours of irrigation given (X16), Total amount of chemical fertilizer applied (X17), Count of livestock (X18) , On farm Income (X19), Communication Variable(X20), Consumption of coal, fuel wood, fire wood etc. (X21)). In the same way the following consequent variables Rice stubble height left in the Field (Y2),Crop residue left in the field (Y3),Ecosystem output (Y7)can operate combined and concomitantly, while they bestow a trend for clandestine selection of right side variable (Education (X2), Size of holding (X5),Number of fragment (X6), Average size of fragment(X7), Size of homestead land (X9), Land under irrigation (X12), Rainfed area (X13),Marketed surplus (X11) , Production of cowdung/poultry liter /goat/biogas slurry (X22)). For this village these function of aspects can be dealt with together while other stewardship function Application of organic manure (Y1), Soil conservation level (Y4), Crop Biodiversity conservation (Y5), Climate resilient technology (Y6) can have a unique combination as well. Both of this group of criterion variables, while interacting with each other, can operationally tune to the respective system variables in a selective manner.

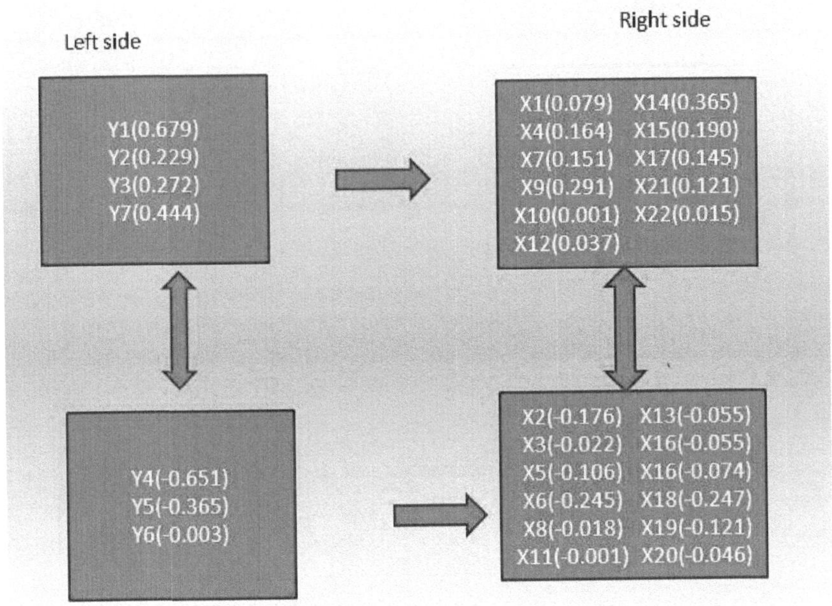

Fig 6.B.30 CANONICAL CORRELATION ANALYSIS: VILLAGE 2 (Bhabanipur)

Results:

The paradigm on canonical correlation analysis depicts that the four consequent variables viz, Application of organic manure (Y1), Rice stubble height left in the Field (Y2), Crop residue left in the field (Y3), Ecosystem output (Y7)show an affinity to move or interact together in the left side echelon, while they bestow a trend for clandestine selection of right side variable (Age (X1), Size of family members (X4), Average size of fragment(X7), Size of homestead land(X9), Yield of Crop(X10), Land under irrigation (X12), Cropping intensity(X14) , Total amount of seed used (X15), Total amount of chemical fertilizer applied (X17), Consumption of coal, fuel wood, fire wood etc.(X21), Production of cowdung/poultry liter /goat/biogas slurry(X22)). In the same way the following consequent variables Soil conservation level (Y4), Crop Biodiversity conservation (Y5), Climate resilient technology (Y6)can operate combined and concomitantly, while they bestow a trend for clandestine selection of right side variable (Education (X2), Functional Education (X3), Size of holding (X5), Number of fragment (X6), Mean distance between two fragments(X8), Marketed surplus(X11), Rainfed area (X13), Total hours of irrigation given (X16), Count of livestock (X18), On farm Income (X19), Communication Variable(X20)) . For these three villages these function of aspects can be dealt with together while other stewardship function Application of organic manure (Y1), Rice stubble height left in the Field (Y2), Crop residue left in the field (Y3), Ecosystem output (Y7) can have a unique combination as well. Both of this group of criterion variables, while interacting with each other, can operationally tune to the respective system variables in a selective manner.

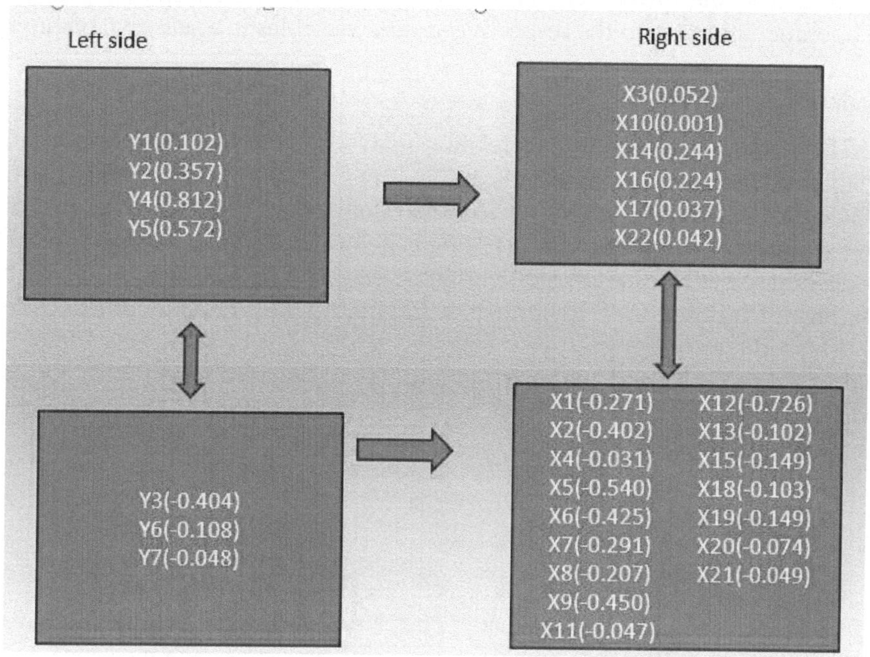

Fig 6.B.31 CANONICAL CORRELATION ANALYIS: VILLAGE 3 (Satypul)

Results :

The paradigm on canonical correlation analysis depicts that the four consequent variables viz, Application of organic manure (Y1), Rice stubble height left in the Field (Y2), Crop residue left in the field (Y3), Climate resilient technology (Y6), Ecosystem output (Y7) have shown an affinity to move or interact together in the left side echelon, while they bestow a trend for clandestine selection of right side variable (Functional Education (X3),Yield of Crop(X10), Cropping intensity(X14) , Total hours of irrigation given (X16), Total amount of chemical fertilizer applied (X17), Production of cowdung/poultry liter /goat/biogas slurry(X22)). In the same way the following consequent variables Crop residue left in the field (Y3), Climate resilient technology (Y6) ,Ecosystem output (Y7), can operate in combination and concomitantly, while they bestow a trend for a clandestine selection of right side variable Age (X1), Education (X2), Count of livestock (X18), On farm Income (X19), Communication Variable(X20), Consumption of coal, fuel wood, fire wood etc.(X21), Marketed surplus(X11), Land under irrigation (X12), Rainfed area (X13),Size of family members (X4), Size of holding (X5),Number of fragment (X6), Average size of fragment(X7), Mean distance between two fragments(X8), Size of homestead land(X9). For these three villages these function of aspects can be dealt with together while other stewardship function Application of organic manure (Y1), Rice stubble height left in the Field (Y2), Crop residue left in the field (Y3), Climate resilient technology (Y6), Ecosystem output (Y7)can have a unique combination as well. Both of this group of criterion variables, while interacting with each other, can operationally tune to the respective system variables in a selective manner.

Revelation:

The paradigm on canonical correlation analysis depicts that the four consequent variables viz, Application of organic manure (Y1),), Crop residue left in the field (Y3),Climate resilient technology (Y6),Ecosystem output (Y7).show an affinity to move or interact together in the left side echelon, while they bestow a trend for clandestine selection of right side variable ((Age (X1), Size of family members (X4), Size of holding (X5),Number of fragment (X6), Functional Education (X3), Mean distance between two fragments(X8), Size of homestead land(X9), Yield of Crop(X10), Land under irrigation (X12), Rainfed area (X13), On farm Income (X19), Consumption of coal, fuel wood, fire wood etc.(X21), Production of cowdung/,poultry liter /goat/biogas slurry(X22)).

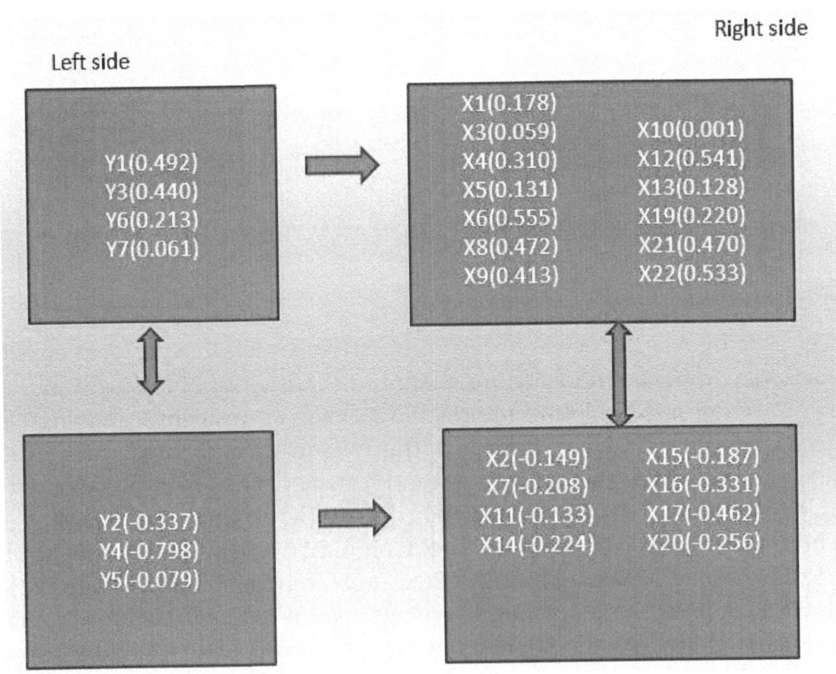

Fig 6.B.32 Canonical Correlation Analysis: Pooled Sample

In the same way the following consequent variables Rice stubble height left in the Field (Y2), Soil conservation level (Y4), Crop Biodiversity conservation (Y5), can operate combined and concomitantly, while they bestow a trend for clandestine selection of right side variable (Education (X2), Average size of fragment (X7), Marketed surplus (X11), Cropping intensity (X14), Total amount of seed used (X15), Total hours of irrigation given (X16), Total amount of chemical fertilizer applied (X17), Count of livestock (X18), Communication Variable (X20). For these three villages, these functional aspects can be dealt with together while other stewardship function Application of organic manure (Y1), Crop residue left in the field (Y3), Climate resilient technology (Y6), and Ecosystem output (Y7) can have a unique combination as well. Both of this group of criterion variables, while interacting with each other, can operationally tune to the respective system variables in a selective manner.

This strategic integration may invite a micro level policy for these villages and be extrapolated to the villages, having more or less, similar social ecology and functions.

8 CHAPTER

CONCLUSION

Natural resource management, conservation and stewardship are nothing but the socio ecological continuity to integrate nature with humanity and humanity with nature. Due to depletive nature of chemicalized agriculture, the ecological resilience has gone down to an alarmingly inelastic level. So, the present study has been atuned and configured to isolate, elucidate and characterize the variables operating in and responsible for ecological functions and deliverables of small holons of farms followed and practiced by the farmers.

From conservation to stewardship, the range and spectrum, can well be tracked when the relational and interactional analysis have been carried out amongst and between the stewardship variables with those of system variables to conclude as to how and where these variables are entering the niche of complex, ecological functions.

The variables with predominant impacts, some few can be taken for interpretation e.g., marketed surplus, yield, consumption of green energy sources, land under irrigation, count of livestock have come up as empirical determinants for shaping and measuring stewardship functions as a whole.

Indian agriculture today is facing the problem of land fragmentation having consequential impact on energy, resource, and cost as well as manpower management. The entitlement process to land ownership here in India and in compliance with population pressure has been harshly instrumental in making land resources operationally disturbing and economically fragile but certain advantages are also there, reflected through a positive correlation between organic manure application, crop residue left in the field and number of fragments for an owner of small or medium holding having more number of fragments sprawling over various soil topography has been 'provocative' to go for opting stewardship functions as an important measure to keep the soil health organically secure. The other variables like count of livestock, as well as land under irrigation have demanded a dyad of affect for combining ecological functions with productive agricultural functions.

The canonical correlation analyses have nicely depicted a gregarious pattern by combining the predicted characters viz; Application of organic manure, Crop residues left in the field, Ecosystem output, Climate resilient technology which have been enough enthusiastic to opt for a combined as well as concerted functions towards accomplishing successful stewardship functions. On the other hand Rice stubble height left in the field, Soil conservation level, Crop biodiversity conservation have developed a symphony kinship as against a selected system variables.

While we will be having a micro level policy intervention, then we have to keep it in our mind that even if stewardship functions is a composite propositions, it has to have a kind of sub layering by accommodating different predicted characters into two or more homogeneous echelon.

However, some important conclusive remarks can further be recorded for incubating and internalizing the extrapolation of research based findings into a system driven operations to spearhead conservation and stewardship functions as a continuum, not as a discrete phenomenon.

- Land fragmentation has been an important determinant to go for stewardship functions, extended up to efficient management of input resources and energy.
- Production and productivity including marketability of the produces have impacted on conservatory stewardship decision of the farmers.
- The stewardship functions are followed either in cognitive, non cognitive or pseudo –cognitive ways which merits a launching of mass awareness campaign to socialize the stewardship in the form of community movement.
- Irrigation based agro ecosystem vis à vis social ecology, despite of inviting problems of chemicalizations and occupational hazards, the practicing farmers have also started eliciting dents for resource conservation and stewardship functions.
- Indigenous knowledge and traditional wisdom have well being synchronized with some promising stewardship functions like Macha (Bamboo structure) in pointed guard, green mulching, land shaping, and so on. Maintenance of traditional and newly entered tree and crop diversity.
- Some marker variables like stubble, crop residues, organic manure, crop biodiversity can well be treated as critical factors for ushering stewardship functions amongst the farmers. A cross cultural and cross entrepreneurial socialization of such factors as mentioned can generate a desired impact.
- For some regression analysis, the R2 values run at a low ebb, the reasons may be stewardship as very nascent concept not perceived well by the practicing farmers, so stimulate questions in certain cases have not been able extract objective responses, just to improve the quantification process subsequently.

9 CHAPTER

FUTURE SCOPE OF THE STUDY

The present study has well germinated the needs of future researches and these are:

- Relegating climate change with stewardship function.
- Methodological innovations for the furtherance of advance researches.
- The socio ecological viability and efficacy of community gene bank, water auditing, disease pest scouting, scope of ICT intervention in stewardship , the ecological proficiency of traditional practices, redeeming and reengineering and soil water and biodiversity management can offer a new scope of future researches.
- The policy issues and paradigming of research interventions can go hand in hand to offer a new goner of research in extension science.
- The eminent research agenda under the aegis of stewardship and resource monitoring can incorporate energy metabolism, farm function entropy, frame working of micro ecology, geo spatial dynamics of social ecology, stress and apocalyptic impact of urbanization the magnitude and dimensions, gender and ethnic issues of forest ecology, community mobilization, auditing , empowerment and entitlement can also offer a huge future scope of extension researches.

Once our good Earth has woombed the whole of the civilization, the embryonic human society has been up scaled to a humongous global civilization but by this time our greed not the need have destroyed an inflicted both the anatomy and physiology of mother earth. Time is just ticking out and shall we remain thoughtless and mindless?

Recommendations

- Stewardship functions and conservation education should be incorporated in the present course and curriculum in order to generate basic and advance knowledge.

- The farm innovation like community gene bank, peoples biodiversity registrar, water Panchayet, folk gene consortia, soil conservation task force have to be created with necessary legal as well as statutory provision.
- Both up scaling and backstopping of technologies are essential for making stewardship function a sustainable proposition and practices.
- Agricultural researches along with extension researches should go for frame working, resource monitoring and stewardship functions therewith.
- Baseline surveys and bench marking information are essentials as to what and how the farm energy, fertilizer and agri chemicals, water and biodiversities are exploited, regenerated or eroded to frame up a community level policy with a proper justification for stewardship functions.
- Auditing, costing and Financing after stewardship function should also be matched or complied with ecological deliverables generated by respective stewardship functions at present and in posterity.
- High end statistical packages are to be applied for efficient and high end monitoring of erosion depletion or retrieval of natural resources. 3 dimensional modeling, ARIMA, Time series analysis, sensitivity analysis can help segregating optimizing and customizing the stewardship functions as applicable in a big volumes of meta data systems with extractable technical information there from.

BIBLIOGRAPHY

Acharjee S K, Ghosh U, Biswas A and Murah A.K., 2004, A study on the perceived benefits of farm forestry in terms of socio economic and motivational correlates in Coochbehar district of West Bengal, The Indian Foretser, Vol.130, No.3,March 2004.

Adger, W.N. 2006. Vulnerability. Global Environmental Change 16:268 261.

Adger, W.N., N.W. Arnell, and E.L. Tompkins. 2005. Successful adaptation to climate change across scales. Global Environmental Change 15:57 86.

Adhikary, M.M., K. Pradhan, and S. Mishra. 2013. Analysing and ensuring persuasive extension with respect to preventive innovation for sustainable agriculture. Journal of Crop and Weed 9(1):56 68. West Bengal, India: Crop and Weed Science Society.

Agricultural Statistics at a Glance 2014, Government of India, Ministry of Agriculture, Department of Agriculture and Cooperation, Directorate of Economics and Statistics, Oxford University Press.

Agriculture census 2010 11 and West Bengal development report 2010.Agricultural Statistics at a Glance, 2014.

Ahnstrom, J., J. Hockert, H.L. Bergea, C.A. Francis, P. Skelton et al. 2009. Farmers and nature conservation: what is known about attitudes, context factors and actions affecting conservation? Renewable Agriculture and Food Systems 24(1):38 47. Cambridge, UK: Cambridge University Press.

Allen, and Patricia. 1993. Connecting the Social and the Ecological in Sustainable Agriculture. In Food for the Future, ed. Patricia Allen. New York: John Wiley & Sons.

Allison, E.H., B.D. Ratner, B. Asgard, R. Willmann, R. Pomeroy et al. 2012. Rights based fisheries governance: from fishing rights to human rights. Fish and Fisheries13(1):14 29. Oxford, UK: Wiley Blackwell.

Ashraf A. Mohammed and Lokanadan Subbalakshmi, 2017,A Review of Rice Landracesin India and its Inherent Medicinal Values The Nutritive Food Values for Future,International Journal of Current Microbiology and Applied Sciences, ISSN: 2319 7706 Volume 6 Number 12 (2017) pp. 348 354.

Atari, D.O.A., E.K. Yiridoe, S. Smale, and P.N. Duinker. 2009. What motivates farmers to participate in the Nova Scotia environmental farm plan program? Evidence and environmental policy implications. Journal of Environmental Management 90(2):1269 1279. Amsterdam, Netherlands: Elsevier.

Bardsley, D.K., and A.M. Bardsley. 2014. Organising for socio ecological resilience: the roles of the mountain farmer cooperative Genossenschaft Gran Alpin in Graubunden, Switzerland. Ecological Economics 98:11 21. Oxford, UK: Elsevier Ltd.

Beer, J., M. Ibrahim, and A. Schlonvoigt. 2000. Timber production in tropical agroforestry systems of Latin America. XXI IUFRO World Congress.

Berkes, F., J. Colding, and C. Folke, editors. 2003. Navigating Social Ecological Systems: Building Resilience for complexity and Change. Cambridge University Press, Cambridge.

Berry, W.2005. Renewing husbandary: After mechanization, can modern agriculture reclaim its soul? Crop Science 45:1103 1106

Bhattacharyya M, Acharya S k and Chakraborty S K.,2017,Pollinators Unknown: People's perception of native bees in an agrarian district of West Bengal, India, and its implication in conservation. Tropical Conservation Science, Vol 10: 1 14, sage publication.

Birkeland, P.W. 1999. Soils and Geomorphology. 3rd Edition. Oxford University Press, New York.

Blann, K. 2006. Habitat in agricultural landscapes: how much is enough? A state of the science literature review. West Linn, OR: Defenders of Wildlife.

Brklacich, M.J. 1989. A framework for evaluating the sustainability of food production systems in a changing environment. Dissertation Abstracts International A, Humanities and Social Sciences 50(6):p.1764.

Buck, L.E., J.A. Milder, T.A. Gavin, and I. Mukherjee. 2006. Understanding ecoagriculture: a framework for measuring landscape performance. Ecoagriculture discussion paper number2.

Buck, L.E., T.A. Gavin, D.R. Lee, and N.T. Uphoff. 2004. Ecoagriculture: a review and assessment of its scientific foundations. Ithaca, NY: Cornell University.

Buck, L.E., T.A. Gavin, N.T. Uphoff, and D.R. Lee. 2007. Scientific assessment of ecoagriculture systems. In Farming with nature: the science and practice of ecoagriculture (eds S. J. Scherr & J. A. McNeely). Washington, DC:Island Press.

Bunch, R. 2000. Reasons for resiliency: toward a sustainable recovery after Hurricane Mitch. Oklahoma City, OK:World Neighbours.

Cairns, M., and D.P. Garrity. 1999 Improving shifting cultivation in Southeast Asia by building on indigenous fallow management strategies. Agroforesty Syst. 47:37– 48.(doi:10.1023/A:1006248104991)

Campbell, A. 2008. Food, energy, water: conflicting insecurities (and the rare win wins offered by soil stewardship). Journal of Soil and Water Conservation Ankeny. 63(5):149A 151A. Ankeny, USA: Soil and Water Conservation Society.

Carruthers, D.V. 2001. The politics and ecology of indigenous folk art in Mexico. Human Organization 60(4):356 366. Oklahoma City, USA: Society for Applied Anthropology

Cassman, K.G. 1999. Ecological intensification of cereal production systems: Yield potential, soil quality, and precision agriculture. Proceedings of the National Academy of Sciences 96:5952 5959.

Cassman, K.G., A. Dobermann, D.T. Walters, and H. Yang. 2003. Meeting cereal demand while protecting natural resources and improving environmental quality. Annaual Review of Environment and Resources 2:316 351.

Cattaneo, M.G. et al. 2006. Farm scale evaluation of the impacts of transgenic cotton on biodiversity, pesticide use and yield. Proc. Natl Acad. Sci. USA 103:7571–7576. (doi:10.1073/pnas.0508312103).

Chambers, R., and G. Conway. 1991. Sustainable Rural Livelihoods: Practical Conceptsfor the 21st century. IDS, Brighton.

Chapin, F.S., III, M. Hotel, S.R. Carpenter, J. Lubchenco, B. Walker et al. 2006a.

Building resilience and adaptation to manage arctic change. Ambio 35:19 202.

Chapin, F.S., III, P.A. Matson, and H.A. Mooney.2002. Principles of Terrestrial Ecosystem Ecolog. Springer Verlag, New York.

Chapin,F S.III, Carpenter R.Stephen, Kofinas Gary P., Folke Carl,Abel N, Clark,W C., Olsson, P, Smith D. Mark S ,et. al.,Ecosystem stewardship: sustainability strategies for a rapidly changing planet.2009 Elsevier Ltd. All rights reserved. doi:10.1016/j.tree.2009.10.008 Available online 16 November 2009.

Chapin, F.S.I., S.R. Carpenter, G.P. Kofinas, C. Folke, N. Abel, et al. 2010. Ecosystem stewardship: sustainability strategies for a rapidly changing planet. Trends in Ecology and Evolution 25(4):241 249. Oxford, UK: Elsevier Ltd.

Clay, J. 2004. World agriculture and the environment: acommodity by commodity guideto impacts and practices. Washington, DC: Island Press.

Colman, D. 1994. Ethics and externalities: agricultural stewardship and other behaviour: presidential address. Journal of Agricultural Economics 45(3):299 311.

Convention on Biological Diversity 2002 Text of the convention on biological diversity. See www.biodiv.org.Convention on Biological Diversity 2006 Decision VI/9: Global strategy for plant conservation.

Conway, G. 1997. The Doubly Green Revolution: Food for All in the 21st Century. Cornell University Press, Ithaca.

Cooke, R.J., and J.R. Park. 2001. The use of a decision support tool to assess the financial viability of selected Countryside Stewardship options. Farm Management11(3):193 204. Reading, UK: Institute of Agricultural Management, Farm Management Unit, Reading University.

Coombe, R.I. 1996 Watershed protection: a better way. In Environmental enhancement through agriculture. Proc. Conf. held in Boston, 15–17 November, 1995 (ed. W. Lockeretz), Organized by the Tufts University School of Nutrition Science, the American Farmland Trust, and the Henry A. Wallace Institute of Alternative Agriculture, pp. 25–34.Medford, MA: Center for Agriculture, Food, and Environment, Tufts University.

Costanza, Robert., Ralph d'Arge, Rudolf de Groot, Stephen Farber, et al. 1997. The Value of the World's Ecosystems Services and Natural Capital. Nature 387: 253 260.

Daily, and Gretchen, ed. 1997. Nature's Services: Societal Dependence on Natural Ecosystems. Washington, D.C: Island Press.

Davies, B.B., and I.D. Hodge. 2007. Exploring environmental perspectives in lowland agriculture: A Q methodology study in East Anglia, UK. Ecological Economics 61(2/3):323 333. Oxford, UK: Elsevier.

Deasy, C., J.N. Quinton, M. Silgram, A.P. Bailey, B. Jackson, and C.J. Stevens. 2010. Contributing understanding of mitigation options for phosphorus and sediment to a review of the efficacy of contemporary agricultural stewardship measures. Agricultural Systems 103(2):105 109. Oxford, UK: Elsevier Ltd.

De Haan, L. R., Cox, T. S., Van Tassal, D. L., & Glover, J. D.2007 Perennial grains. In Farming with nature: the science and practice of ecoagriculture (eds S. J. Scherr& J. A. McNeely). Washington, DC: Island Press

Delate, K. 2002. Using an agroecological approach to farming systems research. HortTechnology12(3):345 354. Alexandria, USA: American Society for Horticultural Science.

Dia, S., J. Fargione, F.S. Chapin, III, and D. Tilman. 2006. Biodiversity loss threatens human well being. Plant Library of Science (PLoS) 4:1300 1305.

Diamond, J. 2005. Collapse: How Socities Choose or Fail to Succeed. Viking, New York.

Diaz, S.J., J. Fargione, F.S. Chapin, III, and D. Tilman. 2006. Biodiversity loss threatens human well being. Plant Library of Science (PLoS) 4: 1300 1305.

Dietz, T., E. Ostrom, and P.C. Stern. 2003. The struggle to govern the commons. Science

302:190 1912.

Ditsch, D.C., R.L. Trimble, and J.M. Wade. 1988. Tillage selection: soil stewardship versus financial survival. Special Bulletin, Mississippi Agricultural and Forestry Experiment Station 88(1): 5 8.

Douglass, and K. Gordon. 1984. Agricultural Sustainability in a Changing World Order.

Boulder, Colorado: Westview Press.

El Baltagy, A. 1996. West Asia and North Africa: a regional vision. West Asia and NorthAfrica: a regional vision 16 pp. Aleppo. Syria: International Center for Agricultural Research in the Dry Areas.

Ellefson, P.V., and M.A. Kilgore. 2010. United States wood based industry: a review of structure and organization. Staff Paper Series Department of Forest Resources, University of Minnesota (206):238 pp. St. Paul, USA: Department of Forest Resources, University of Minnesota.

Elmqvist, T., C. Folke, M. Nystrom, G. Peterson, J. Bengtsson, et al. 2003. Response diversity, ecosystem change, and resilience. Frontiers in Ecology and the Environment 1:488 494.

Falcon, W.P., and R.L. Naylor. 2005. Rethinking food security for the 21st century.

American Journal of Agricultural Economics 87:1113 1127.

Farnham, UK: British Crop Protection Council.

Fayram, A.H., D.A. Schenborn, J.M. Hennessy, N.A. Nate, P.J. Schmalz. 2009. Exploring the conflict between broad scale and local inland fisheries management: the risks to agency credibility. Fisheries Bethesda 34(5):232 236. PB: Bethesda, USA: American Fisheries Society.

Fischer, J., A.D. Manning, W. Steffen, D.B. Rose, K. Daniell, et al. 2007. Mind the sustainability gap. Trends in Ecology and Evolution 22:621 624.

Fischer, J., R. Dyball, I. Fazey, C. Gross, S. Dovers, P.R. Ehrlich, R.J. Brulle, C. Christensen and R.J. Borden. 2012. Human behavior and sustainability. Frontiers in Ecology and the Environment 10(3):153 160. PB: Washington, USA: Ecological Society of America. URL: http://www.esajournals.org/doi/full/10.1890/110079

Foley, J.A., R. DeFries, G.P. Asner, C. Barford, G. Bonan, et al. 2005. Global consequences of land use. Science 309:570 574.

Folke, C. 2003. Freshwater for resilience: A shift in thinking. Philisophical Transcation of the Royal Society of London, Series B 35:2027 2036.

Folke, C. 2006. Resilience: The emergence of a perspective for social – ecological system analysis. Global Environmental Change 16:253 267.

Folke, C., J. Codling, and F. Brerkes. 2003. Synthesis: Building resilience and adaptive capacity in social ecological system. Pages 352 387 in F. Berkes, J. Clding, and

C. Folke, editors. Navigating Social Ecological Systems: Building Resilience for Complexity and Change. Cambridge University Press, Cambridge.

Forman, R. T. 1995 Land mosaic: the ecology of landscapes and regions. Cambridge, UK: Cambridge University Press.

Fox, G. 1994. The structure of ownership and prosperity. Canadian Journal of Agricultural Economics 42(4):511 524.

Franklin, S.B., J.A. Kupfer, J.W. Grubaugh, and M.L. Kennedy. 2004. Biotic diversity ofNatchez Trace State Forest, Western Tennessee. Environmental Monitoring and Assessment 93(1/3):31 54. Dordrecht, Netherlands: Kluwer Academic Publishers.

Gilmore, D.W. 1997. Ecosystem management a needs driven, resource use philosophy.

Forestry Chronicle 73(5):560 564.

Graham, D A; Bonta, M A; Ulloa, R. 2011. Cycad conservation, peasant subsistence, and the military coup in Honduras. Society and Natural Resources 24(2):193

200. London, UK: Routledge, Taylor & Francis.

Cycad conservation, peasant subsistence, and the military coup in Honduras

Grigg N.S. 2008. Integrated water resources management: balancing views and improving practice. Water International. 33(3):279 292. Carbondale, USA: International Water Resources Association (IWRA).

Fehrenbach, H. 2011. How bioenergy related water impacts are considered by certification schemes. Biofuels, Bioproducts and Biorefining 5(4):464 473. Chichester, UK: Wiley Blackwell.

Grossi, R.E. 1993. A green evolution: retooling agricultural policy for greater sustainability.Journal of Soil and Water Conservation 48(4):285 288.

Gunderson, L.H., and C.S. Holling, editors. 2002. Panarchy: Understanding transformations in Human and Natural Systems. Island Press, Washington.

Hancock, D.W. 2012. Grazing schools improve the sustainability of pasture based livestock enterprises. Journal of Extension50(5): 5IAW5. West Lafayette, USA: Extension Journal Inc.

Hartmann, M., C.G. Howes, D. VanInsberghe, H. Yu, D. Bachar, et al. 2012. Significant and persistent impact of timber harvesting on soil microbial communities in northern coniferous forests. ISME Journal 6(12): 2199 2218. London, UK: Nature Publishing Group.

Harvey, C.A., J.C. Sanez, and J. Montero. 2005. Conservacion de la biodiversidad en paisajes fragmentados y rurales de Mesoamerica? Que hemos aprendido y que todavı´a necesitamos conocer? Turrialba, Costa Rica: CATIE. Draft.

Hoag, D.L.K and R.B. Boone, and C.M.H. Keske. 2011. The cost for agriculture to coexist with wildlife in Colorado. Human Dimensions of Wildlife 16(5):318 329 Philadelphia, USA: Taylor & Francis, Inc.

Hole, D. G., A.J. Perkins, J.D. Wilson, I.H. Alexander, I. H., Grice, P. V. & Evans, A. D.2005 Does organic farming benefit biodiversity? Biol. Conserv. 122:113–130. (doi:10. 1016/j.biocon.2004.07.018).

Holling, C.S. 1973. Resilience and stability of ecological systems. Annual Review of Ecology and Systematics 4:1 23.

Holling, C.S., and L.H. Gunderson. 2002. Resilience and adaptive cycles. Pages 25 62 in

L.H. Gunderson and C.S. Holling, editors. Panarchy: Understanding Transformations in Human and Natural Systems. Island Press, Washington.

Holmes, M.C.C., and W. Jampijinpa. 2013. Law for country: the structure of Warlpiri ecological knowledge and its application to natural resource management and ecosystem stewardship. SO: Ecology and Society 18(3): Art. 19, Waterloo, Canada: Resilience Alliance.

Holsman, R.H. 2000. Goodwill hunting? Exploring the role of hunters as ecosystem stewards. Wildlife Society Bulletin 28(4): 808 816. Bethesda, USA: Wildlife Society.

Hoogheem, T.J. 1997. Global product stewardship water quality protection.1997 Brighton crop protection conference: weeds. Proceedings of an international conference, Brighton, UK, 17 20 November 1997. 1997; (Volume 3): 1085 1090.

IITA (International Institute of Tropical Agriculture). 2001. Delivering the African agricultural development potential: theIITA vision (Strategic Plan 2001–2010). Ibadan, Nigeria:IITA.

Ikerd, and E. John. 1993. The Need for a Systems Approach to Sustainable Agriculture.

Agriculture, Ecosystems and Environment 46:147 160.

IPCC (Intergovermental Panek on Climate Change). 2007a. Climate change 200: The Physical Science Basis, Contribution of Working Group I to the Fourth Assessment Report of the Intergovernmental Panel on Climate Change. Cambridge University Press, Cambridge.

Jackson, D.L., and L.L. Jackson. 2002. The farm as habitat: reconnecting food systems with ecosystems.Washington, DC: Island Press.

Jackson, J.B.C. 2001. What was natural in the coastal oceans? Proceedins of the National Academy of Sciences 9:5411 5418.

Jackson, L., K. Bawa, U. Pascual, and C. Perrings, C. (eds) 2005. AgroBiodiversity: a new science agenda for biodiversity insupport of sustainable agroecosystems. DIVERSITASreport no. 4. Davis, CA: DIVERSITAS.

Jackson, W., and L.L. Jackson. 1999. Developing high seed yielding perennial polycultures as a mimic of mid grass prairie. In Agriculture as a mimic of natural ecosystems (edsR. LeFroy, E. C. Lefroy, R. J. Hobbs, M. H. O'Connor &J. S. Pate), pp. 1–37. Dordrecht, The Netherlands: Kluwer Academic Publishers.

Janssen, M.A., M.L. Schoon, W. Ke, and K. Borner. 2006. Schlorly networks on resilience, vulnerability and adaptation within the human dimensions of global environmental change. Global Environmental Changes 16:240 252.

Jat, R.A., K.L. Sahrawat, A.H. Kassam, T. Friedrich. 2014. Conservation agriculture for sustainable and resilient agriculture: global status, prospects and challenges.

Conservation agriculture: global prospects and challenges1 25. Wallingford, UK: CABI.

Kang, M.S., and S.S. Banga. 2013. Global agriculture and climate change. Journal of Crop Improvement 27(6):667 692. Philadelphia, USA: Taylor & Francis.

Kay, P., A.C. Edwards, and M. Foulger. 2009. A review of the efficacy of contemporary agricultural stewardship measures for ameliorating water pollution problems of key concern to the UK water industry. Agricultural Systems 99(2/3):67 75. Oxford, UK: Elsevier.

Kay, P., R. Grayson, M. Phillips, K. Stanley, A. Dodsworth, et al. 2012. The effectiveness of agricultural stewardship for improving water quality at the catchment scale: experiences from an NVZ and ECSFDI watershed. Journal of Hydrology Amsterdam 422/423:10 16. Oxford, UK: Elsevier Ltd.

Keith M. a,n, Schmidt Olaf b, Barry J. McMahon b. 2016 Soil stewardship as an exus between Ecosystem Services and One Health Aidan, Elsevier, A Centre for Ecology & Hydrology, Lancaster Environment Centre, Library Avenue, Bailrigg,Lancaster LA14AP,UK b UCD School of Agriculture & Food Science, University College Dublin, Belfield, Dublin4, Ireland, Ecosystem Services 17(2016) 40–42, Contents lists available at Science Direct

Kleijn, D., A.R. Baquero, Y. Clough, M. Dı´az, J. De Esteban, and F. Ferna´ndez. 2006. Mixed biodiversity benefits of agri environment schemes in five European countries. Ecol. Lett. 9:243–254. (doi:10.1111/j.1461 0248.2005.00869.x)

Klein, D.J. and W.J. Sutherland. 2003. How effective are European agri environment schemes in conserving and promoting biodiversity? Appl. Ecol 40:947–969. (doi:10.1111/j.1365 2664.2003.00868.x)

Kloppenburg, Jack, Hendrickson, John, and G.W. Stevenson. 1996. Coming Into the Foodshed. In Rooted in the Land, ed. W. Vitek and W. Jackson, 177 186. New Haven and London: Yale University Press.

Kloppenburg, Jack, Lezberg, Sharon, DeMaster, et al. 2000. Tasting food, Tasting Sustainability: Defining the Attributes of an Alternative Food system with Competent, Ordinary People," Human Organization 59(2):177 186.

Kogan, M. 1998 Integrated pest management: historical perspectives and contemporary developments. Annual Review of Entomology 43:243–270. (doi:10.1146/annurev.ento.43.1.243)

Krasny, M.E., and K.G. Tidball. 2012. Civic ecology: a pathway for Earth Stewardship in cities.Frontiers in Ecology and the Environment10(5):267 273. Washington, USA: Ecological Society of America.

Kremen, C., M.N. Williams, and R.W. Thorp. 2002. Thorp crop pollination from native bees at risk from agricultural intensification. Proc. Natl Acad. Sci. USA 99:812– 816. (doi:10.1073/pnas.262413599).

Lahiri Dutt, K. 2014. Experiencing and coping with change: women headed farming households in the Eastern Gangetic Plains. ACIAR Technical Reports Series

(83):66 pp. Canberra, Australia: Australian Centre for International Agricultural Research (ACIAR). URL: http://www.aciar.gov.au/taxonomy/term/15

Lai PoHsin, M.G. Sorice, S.K. Nepal, and Cheng ChiaKuen. 2009. Integrating social marketing into sustainable resource management at Padre Island National Seashore: an attitude based segmentation approach. Environmental Management 43(6):985 998. New York, USA: Springer Verlag.

Lambin, E.F. 2007. The Middle Path: Avoiding Environmental Catastrophe. University of Chicago Press, Chicago.

Landers, J.N. 2004. Zero tillage challenges and solutions in Brazil. Proceedings VII World Soybean Conference, IV International Soybean Processing and Utilization Conference, III Congresso Brasileiro de Soja Brazilian Soybean Congress, Foz do Iguassu, PR, Brazil, 29 February to 5 March, 2004; 274 281. Londrina, Brazil: Brazilian Agricultural Research Corporation, National Soybean Research Center

Lange, B., and M. Shepheard. 2014. Changing conceptions of rights to water? An eco socio legal perspective. Journal of Environmental Law 26(2): 215 242. Oxford, UK: Oxford University Press.

Leakey, R. 1999. Agroforestry for biodiversity in farming systems. In Biodiversity in agro ecosystems (edsW.W. Collins& C. O. Qualset), pp. 127–145. New York, NY: CRCPress.

Leakey, R.R.B. and Z. Tchoundjeu. 2001 Diversification of tree crops: domestication of companion crops for poverty reduction and environmental Services. Exp. Agr. 37:279–296. (doi:10.1017/S0014479701003015)

Leaman, D.J. 2006. Sustainable wild collection of medicinal and aromatic plants: development of an international standard. Medicinal and aromatic plants: agricultural, commercial, ecological, legal, pharmacological and social aspects 97 107. Heidelberg, Germany: Springer Verlag GmbH.

Lefroy, R. et al. 1999. Agriculture as a mimic of natural ecosystems. Current plant science and biotechnology in agriculture, vol. 37. Dordrecht, The Netherlands: Kluwer Academic Publishers.

Leopold, A. 1949. A Sand County Almanac. Oxford university Press, Oxford.

Lertzman, K. 2009. The paradigm of management, management systems, and resource stewardship. Journal of Ethnobiology 29(2):339 358. Denton, USA: Society of Ethnobiology.

Letey, J., R.E. Sojka, D.R. Upchurch, D.K. Cassel, K.R. Olson, et al. 2003. Deficiencies in the soil quality concept and its application. Journal of Soil and Water Conservation Ankeny 58(4):180 187. Ankeny, USA: Soil and Water Conservation Society.

Lichtenberg, E. 2000. Agriculture and the environment. Working Paper Department of Agricultural and Resource Economics, University of Maryland (00/15):111 pp. PCollege Park, USA: University of Maryland.

Lindenschmidt, K.E., G. Syrenne, and R. Harrison. 2010. Measuring ice thicknesses along the Red River in Canada using RADARSAT 2 satellite imagery. Journal ofWater Resource and Protection 2(11): 923 933. Irvine, USA: Scientific Research Publishing.

L M;N. Liggett, D. Havens, W.C. Anderson, L.K. Stark. 1994. Agricultural stewardship:building partnerships through a focus group process. An overview of the WSU Skagit County Water Quality and Nitrate Leaching Focus Group...a model for building partnerships between diverse interest groups. Extension BulletinCooperative Extension, College of Agriculture and Home Economics, Washington State University (EB1788):19. Pullman, Washington, USA.

Lobell, D.B., and J.I. Orti Monasterio. 2008. Satellite monitoring of yield responses to irrigation practices across thousands of fields. Agronomy Journal 10:1005 1012.

Lowthe Thomas, S.C., A. Clarke, S.J. Clapham, M. Morgan, T. Semere, et al. 2008. Reed canary grass a versatile energy crop for Wales? Aspects of Applied Biology (90):81 86. Wellesbourne, UK: Association of Applied Biologists.

Lu, F., and C. Wirth. 2011. Conservation perceptions, common property, and cultural polarization among the Waorani of Ecuador's Amazon. Human Organization70(3):233 243. Oklahoma City, USA: Society for Applied Anthropology.

Mansuri, G., and V. Rao. 2004. Community based (and – driven) development: A critical review. The World Bank Research Observer 19(1):1 39.

Mason, N. 2013. ODI Working Paper (363): viii + 48 pp. London, UK: Overseas Development Institute.

Matson, P.A., W.J. Parton, A.G. Power, and M.J. Swift. 1997. Agricultural intensification and ecosystem properties. Science 22:504 509.

McBratney, A., D.J. Field, and A. Koch. 2014. The dimensions of soil security.

Geoderma 213:203–213. http://dx.doi.org/10.1016/j.geoderma.2013.08.013

McCarthy, J.J., M.L. Martello, R. Corell, N.E. Selin, S. Fox, et al. 2005. Climate change in the context of multiple stressors and resilience. Pages 945 988 in ACIA,editors. Arctic Climate Impact Assessment. Cambridge University Press.

McIntyre, J. 2010. Measuring agricultural stewardship: risks and rewards. The case for the stewardship index for specialty crops. Journal of Agriculture, Food Systems and Community Development1(1): 19 22. Ithaca, USA: New Leaf Associates.

McNeely, J.A., and S.J. Scherr. 2003. Ecoagriculture: strategies for feeding the world and conserving wild biodiversity. Washington, DC: Island Press.

MEA (Millineum Ecosystem Assessment). 2005a. Ecosystem and Human Being: Current Status and Trends. Island Press, Washington.

MEA (Millineum Ecosystem Assessment). 2005b. Ecosystem and Human Being: Desertification Synthesis. World Resources Institute, Washington.

MEA (Millineum Ecosystem Assessment). 2005c. Ecosystem and Human Being: Scenarios. Island Press, Washington.

MEA (Millineum Ecosystem Assessment). 2005d. Ecosystem and Human Being: Synthesis. Island Press, Washington, D.C.

Millard, E. 2007. Restructuring the supply chain. In Farming with nature: the science andpractice of ecoagriculture (eds S. J.Scherr & J. A. McNeely). Washington, DC: Island Press.

Miller, F.P. 2003. Natural resources management and the second copernican revolution. Journal of Natural Resources and Life Sciences Education32:43 51. Madison, USA: American Society of Agronomy.

Miller, F.P. 2003. Natural resources management and the second copernican revolution. Journal of Natural Resources and Life Sciences Education32:43 51, Madison, USA: American Society of Agronomy.

Mishra S S, Acharjee S K, Chakraborty S K, 2009, Development of tools for assessing conservation categories of siluroid fishes of fresh water and brackish water wetlands of South West Bengal, India. Environ Biol Fish (2009) 84 : 395 407, @ Springer.

Mitchell, D.P.,T.W. Norton, R.I. Grenfell, and P.W. Woodgate. 2007. Barriers to implementation of natural resource management at the landholding level in the intensive agricultural zones of Southern Australia. Australasian Journal of Environmental Management14(2):103 110. Melbourne, Australia: Environment Institute of Australia and New Zealand.

Molden, D., R. Tharme, I. Abdullaev, and R. Puskar. 2007. Irrigation. In Farming with nature: the science and practice of ecoagriculture (eds S. J. Scherr & J. A. McNeely).Washington, DC: Island Press.

Molnar, A. 2003. Forest certification and communities: looking forward to the next decade. Washington, DC: Forest Trends. Molnar, A., Scherr, S. J. & Khare, A. 2004, Who conserves theworld's forests? A new assessment of conservation and investment trends. Washington, DC: Forest Trends.

Mooney, Harold and Paul R. Ehrlich. 1997. Ecosystem Services: A Fragmented History. In Nature's Services: Societal Dependence on Natural Ecosystems, ed. Gretchen

C. Daily, 11 19. Washington, D.C.: Island Press.

Murombedzi, J. 2001. Committees, rights, costs & benefits: natural resource stewardship & community benefits in Zimbabwe's CAMPFIRE programme. African wildlife and livelihoods: the promise and performance of community conservation244 255. Oxford, UK: James Currey Ltd. National Crime Record Bureau Report, 2016.

Naylor, R.L., R.J. Goldburg, J.H. Primavera, N. Kutsky, M.C.M. Beveridge, et al. 2000.

Effect of aquaculture on world fish supplies. Nature 405:101 1024.

Naylor, R.L., W.P. Falcon, R.M Goodman, M.M. Jahn, T. Sengooba, et al. 2004. Biotechnolog in the development world: A case for increased investments in orphan crops. Food Policy 29(1):15 44.

Naylor, RL. 2000. Agriculture and global change. Pages 462 45 in G. Ernst, editor. Eart Systems: Process and Issues. Cambridge University Press, Cambridge.

Nelson, D.R., W.N. Adger, and K. Brown. 2007. Adaptation to environmental change: contributions of a resilience framework. Annual Review of Environment and Resources 32:395 419.

Nierenberg, D. 2005. Happier meals: rethinking the global meat industry. Washington, DC: Worldwatch Institute. Oksman Caldentey, K. M. & Barz, W. H. 2002 Plant biotechnology and transgenic plants. New York, NY: Marcel Dekker.

Odum, E.P. 1989. Ecology and Our Endangered Life Support Systems. Sinauer Associates, Sunderland, MA.

Overloop, P.J. van., I.J. Miltenburg, A.J. Clemmens, and R.J. Strand. 2008. Identification of pool characteristics of irrigation canals. Proceedings of the World Environmental and Water Resources Congress, Honolulu, Hawaii, USA, 12 to 16May 2008, 118. Reston, USA: American Society of Civil Engineers.

Patra M K, Acharjee S K and Chakraborty S K.2005,Conservation categories of siluroid fishes in North East Sundarbans, India, Biodiversity and Conservation (2005) 14: 1863 1876, @ Springer 2005.

Peach, W.J. et al. 2001. Countryside stewardship delivers cirl buntings (Emberiza cirlus) in Devon, UK. Biol. Conserv. 101:361–373. (doi:10.1016/S0006 3207(01) 00083 0)

Pearce, D. 2005. Managing environmental wealth for poverty reduction. Poverty and environment partnership MDG7 Initiative economics.

Peng Shaobing , Huang Jianliang,. Sheehy John E, Laza Rebecca C. , Visperas Romeo

M. , Zhong Xuhua ,Centeno Grace S. ,. Khush Gurdev S, and Cassman Kenneth G.,Rice yields decline with higher night temperature from global warming, Contributed by Gurdev S. Khush, May 27, 2004

www.pnas.org_cgi_doi_10.1073_pnas.0403720101, The National Academy of Sciences of the USA.

Plieninger, T., and C. Bieling. 2013. Resilience based perspectives to guiding high nature value farmland through socioeconomic change. Ecology and Society 18(4): art20,Waterloo, Canada: Resilience Alliance. Plieninger, T., S. Ferranto, L. Huntsinger,

M. Kelly, and C. Getz. 2012. Appreciation, use, and management of biodiversity and ecosystem services in California's working landscapes. Environmental Management 50(3):427 440. New York, USA: Springer

Power, A.G. 1999. Linking ecological sustainability and world food need. Environment, Development, and Sustainability 1:185 196.

Pretty and N. Jules. 1995. Regenerating Agriculture: Policies and Practices for Sustainability and Self Reliance. Washington, D.C.: Joseph Henry Press.

Puste A.M., Gunri S.K., Jana K., Pramanik B. Ray, Acharya S.K., Dasgupta M. and Maity T.K., 2017, Integrating management for land water ecosystem for augmenting productivity, Income, and sustainable livelihood, @ Springer Nature Singapore Pte Ltd. 2017, U.K.De et al. (eds.) , Inequality, poverty and development in India, India Studies in Business and Economics.

Rainforest Alliance. 2006. The certified sustainable products alliance.

Redford, K.H., and B.D. Richter. 1999. Conservation of biodiversity in a world of use.

Conserv. Biol. 13:1246–1256. (doi:10.1046/j.1523 1739.1999.97463.x)

Redman, C.L. 1999. Human Impacts on Ancient Environments. University of Arizona Press, Tucson.

Regan, J.B. 1995. Natural resource stewardship efficacy of Treflan and Sonalan in conservation tillage. Down to Earth Midland 50(2):6 12.

Rhodes, G., P. Henrys, B.C. Thomson, and R.W. Pickup. 2013. Mycobacterium avium subspecies paratuberculosis is widely distributed in British soils and waters: implications for animal and human health. Environ. Microbiol 15:2761–2774.

Ribaudo, M., and Caswell, M.F. 1999. Environmental regulation in agriculture and the adoption of environmental technology. Flexible incentives for the adoption of environmental technologies in agriculture 7 25. Dordrecht, Netherlands: Kluwer Academic Publishers.

Ricketts, T. 2004. Tropical forest fragments enhance pollinator activity in nearby coffee crops. Conserv. Biol. 18:1262–1271. (doi:10.1111/j.1523 1739.2004.00227.x)

Robertson, G.P., and S.M. Swinton. 2005. Recounciling agricultural productivity and environmental integrity: A grand challenge for agriculture. Frontiers in Ecology and the Environmnent 3:38 46.

Rockstrom, J., L. Gordon, C. Folke, M. Falkenmark, and M. Engwall. 1999. Linkages among water vapor flows, food production, and terrestrial ecosystem services. Conservation Ecology 3:5.

Rockstrom, J.,M. Falkenmark, T. Allan, C. Folke, L. Gordon, et al. 2014. The unfolding water drama in the Anthropocene: towards a resilience based perspective on water for global sustainability. Ecohydrology 7(5):1249 1261. Chichester, UK: Wiley Blackwell.URL: http://onlinelibrary.wiley.com/journal/10.1002/(ISSN)1936 0592

Rome, Italy: Food and Agriculture Organization of the United Nations (FAO). 2012. FAO Workshop on International Guidelines for Securing Sustainable Small

ScaleFisheries, Rome, Italy, 7 10 February 2012. FAO Fisheries and Aquaculture Report. 2012; (1004): i vii, 1 44

Rotterdam, Netherlands: A.A. Balkema. 2000. Reclaimed land: erosion control, soils andecology. Reclaimed land: erosion control, soils and ecology 385.

Roy B,Rahman Md. Habibur, and Fardusi Most. Jannatul,2013, Status, Diversity and Traditional Uses of Homestead gardens in North Bangladesh : A Means ofSustainable biodiversity conservation, Hindawi Publishing Coorporation, ISRN Biodiversity, Volume 2013, Article ID 124103, 11 pages

Rozza, J.P.,B.D. Richter, W.M. Larson, T. Redder, K. Vigerstol et al. 2013. Corporate water stewardship: achieving a sustainable balance. Journal of Management and Sustainability 3(4):41 52. Toronto, Canada: Canadian Center of Science and Education. URL:http://www.ccsenet.org/journal/index.php/jms/article/view/28896/18665

Rozzi, R., J.J. Armesto, J.R. Gutierrez, F. Massardo, G.E. Likens, et al. 2012. Integrating ecology and environmental ethics: Earth stewardship in the Southern end of the Americas. SO:BioScience62(3):226 236. Washington, USA: American Institute of Biological Sciences. URL: http://www.bioone.org/doi/full/10.1525/bio.2012.62.3.4

Ruark, G., J. Scott, D. Riemenschneider, and T. Volk. 2006. Perennial crops for bio fuelsand conservation. Presented at 2006 USA Agricultural Outlook Forum—Prospering in Rural America.

Salve Prachi, How many farmers does India really have? The Hindustan Times ,August 11, 2014.

Sanjines, P.B. 2012. Opportunity analysis of payment for ecosystem services: policy design and implementation for coffee agroforestry systems in Costa Rica. Tropical Resources: Bulletin of the Yale Tropical Resources Institute31:56 66. New Haven, USA: Yale School of Forestry and Environmental Studies.

Sara J. Scherr and Jeffrey A. McNeely, Ecoagriculture Partners, The Royal Society Washington, DC 2000, USA., World Conservation Union—IUCN, 1196 Gland, Switzerland Phil. Trans. R. Soc. B (2008) 363, 477–494.

doi:10.1098/rstb.2007.2165. Published online 25 July 2007

Sayer, J.A., and S. Maginnis. (eds) 2005. Forest in landscapes: ecosystem approaches tosustainability. London, UK: Earthscan.

Sayre, N.F., R.R.J. McAllister, B.T. Bestelmeyer, M. Moritz, and M.D. Turner. 2013.

Earth

Stewardship of rangelands: coping with ecological, economic, and political marginality. Frontiers in Ecology and the Environment 11(7):348 354. Washington, USA: Ecological Society of America.URL: http://www.esajournals.org/doi/full/10.1890/120333

Schaller, and Neill. 1993. Sustainable Agriculture and the Environment. Agriculture, Ecosystems and Environment 46: 89 97.

Scherr Sara J., McNeely Jeffrey A. ,Biodiversity conservation and agricultural sustainability: towards a new paradigm of 'ecoagriculture' landscapes Ecoagriculture Partners, Philosophical Transaction of the Royal Society B (2008) 363, 477–494 doi:10.1098/rstb.2007.2165 Washington, DC 2000, USA 2 World Conservation Union—IUCN, 1196 Gland, Switzerland online 25 July 2007

Scherr, and J.A. McNeely. 2017. Biodiversity and agricultural sustainability Phil. Trans.

R. Soc. B (2008) Scherr, S., and A. Meredith. 2005. Ecoagriculture Outcome measures technical workshop report. Washington, DC:Ecoagriculture Partners.

Scherr, S.J., J.C. Milder, and M. Inbar. 2007a. Paying farmers for stewardship. In Farming with nature: the science and practice of ecoagriculture (eds S. J. Scherr& J. A. McNeely).Washington, DC: Island Press

Schroth, G., G. da Fonseca, C. Harvey, C. Gascon, H. Vasconcelos, and M. Izac. 2004. Agroforestry and biodiversity conservation in tropical landscapes. Washington, DC: Island Press

Seidl, R., P.M. Fernandes, T.F. Fonseca, F. Gillet, A.M. Jonsson, et al. 2011. Modelling natural disturbances in forest ecosystems: a review. Ecological Modelling2011; 222(4): 903 924. Oxford, UK: Elsevier Ltd.

Siebert, S.F. 2002. From shade to sun grown perennial cropsin Sulawesi, Indonesia: implication for biodiversity conservation and soil fertility. Biodiv. Conserv. 11:1889–1902.

Smil, V. 2000. Feeding the World: A Challenge for the 21st Century. MIT Press, Cambridge, M.A.

Solow, R.M. 1991. Sustainability: An economist's perspective. Eighteenth J. Seward Johnson Lecture, Marine Policy Center, Woods Holes, MA. Pages 19 187 in R. Dorfman and N.S. Dorfman, editors. Economics of the Environment: Selected Readings. W.W. Norton, New York.

STCP (Sustainable Tree Crops Programme). 2005. SustainableTree Crops Programme: a public private partnership.

Steffen, W.L., A. Sanderson, P.D. Tyson, J. Jager, P.A. Matson, et al. 2004. Global Changes and the Earth System: A Planet Under Pressure. Springer Verlag, New York.

Stokka, G L; Maddock, R; Dahlen, C; Petry, T;Stoltenow, C. 2014. Livestockstewardship: "pregnancy checking, why should we recommend it?". Proceedings of the 46th Annual Conference of the American Association of Bovine Practitioners, Milwaukee, Wisconsin, USA, 19 21 September 2013. 2014; 34 35,Stillwater, USA: American Association of Bovine Practitioners.

Stonehouse, D.P. 2004. Sustainability issues in the agri food sector in Ontario, Canada. Journal of Sustainable Agriculture 23(3):109 124, Binghamton, USA: Food Products Press.

Suding, K.N., S. Lavorel, F.S. Chapin, III, J.H.C. Cornelissen, S. Diaz, et al. 2008. Scaling environmental change through the community level: A trait based response and effect framework for plants. Global Change Biology 14:1125 1140.

Swaminathan, M. S. (ed.) 1994 Ecotechnology and rural employment: a dialogue.

Chennai, India: Macmillan India.

Swift, M.J., A.M.N. Izac, and M. van Noordwijk. 2004. Biodiversity and ecosystem services in agricultural landscapes—are we asking the right questions? Agr. Ecosyst. Environ. 104: 113–134. (doi:10.1016/j.agee. 2004.01.013)

Thompson, J., T. Hodgkin, K. Atta Krah, D. Jarvis, C. Hoogendoorn, and S. Padulosi. 2007. Biodiversity in agroecosystems. In Farming with nature: the science and practice of ecoagriculture (eds S. J. Scherr & J. A. McNeely).Washington, DC: Island Press.

Thompson, P.B. 2011. The ethics of soil: stewardship, motivation, and moral framing. Sustainin soil productivity in response to global climate change: science, policy, and ethic31 4. Chichester, UK: John Wiley & Sons.

Ticknor, W.D. 1996. Arkansas forestry: maintaining the momentum of excellence. SpecialReport Agricultural Experiment Station, Division of Agriculture, University of Arkansas(175): v + 10 pp. Fayetteville, Arkansas, USA:

Tilman, D. et al. 2001. Forecasting agriculturally driven global environmental change.

Science 292:281–284. (doi:10.1126/science.1057544).

Tilman, D., K. Cassman, P. Matson, R. Naylor, and S. Polasky. 2002. Agricultural sustainability and intensive production practices. Nature 41:671 677.

Tomich, T.P., M. van Noordwijk, and D.E. Thomas. 2004. On bridging gaps. Agr.

Ecosyst. Environ. 104:1–3. (doi:10.1016/j.agee.2004.01.002)

Trewavas, A.J. 2001. The population/biodiversity paradox: agriculture efficiency to save wilderness. Plant Physiol. 125:174–179. (doi:10.1104/pp.125.1.174)

Turner, B.L., II, R.E. Kasperson, P.A. Matson, J.J. McCarthy, R. W. Corell, et al. 2003. A framework for vulnerability analysis in sustainability science. Proceedings of the National Academy of Sciences 100:8074 8079.

Turner, B.L., II, W.C. Clark, R.W. Kates, J.S. Richards, J.T. Mathews, et al. editors. 1990. The Earth as Transformed by Human Action: Global and Years. Cambridge University Press, Cambridge.

Turner, N.J., M. Plotkin, and H.V. Kuhnlein. 2013. Global environmental challenges to the integrity of Indigenous Peoples' food systems. Indigenous peoples' food systems and well being: interventions and policies for healthy communities 23 38.

Uphoff, N. et al. (eds) 2006 Biological approaches to sustainable soil systems. Boca Raton, FL: CRC Press.

Vitousek, P.M., H.A. Mooney, J. Lubchenco, and J.M. Melillo. 1997. Human domination of Earth's ecosystems. Science 277:494 499.

Walker, B.H., and D. Salt. 2006. Resilience Thinking: Sustaining Ecosystems and Peoplein a Changing World. Island Press, Washington.

Walker, B.H., and J.A. Mayers. 2004. Thresholds in ecological and socio ecological systems: A developing database. Ecology and Society 9(2):3.

Walker, B.H., C.S. Holling, S.R. Carpenter, and A.P. Kinzig. 2004. Resilience, adaptability and transformability in social ecological systems. Ecology and Society 9(2):5

Walker, B.H., L.H. Gunderson, A.P. Kinzig, C. Folke, S.R. Carpenter, et al. 2006. A handful of heuristics and some propositions for understanding resilience in socio ecological systems. Ecology and Society 11(1):13.

Wander, M.M., and L.E. Drinkwater. 2000. Fostering soil stewardship through soil quality assessment. Applied Soil Ecology 15(1):61 73.

Washington, D.C., USA: National Academy Press. 1988. Understanding agriculture: new directions for education. Understanding agriculture: new directions for educationix + 68pp.

Water Stewardship in Agriculture, WWF Brief. 2015. Together in partnership with IUCN

Weber, T.C., and W.L. Allen. 2010. Beyond on site mitigation: an integrated, multi scaleapproach to environmental mitigation and stewardship for transportation projects.Landscape and Urban Planning 96(4): 240 256. Oxford, UK: Elsevier Ltd.

Weibull, A.C., O. Ostman, and A. Granqvist. 2003. Species richness in agroecosystems: the effect of landscape, habitat and farm management. Biodivers. Conserv. 12:1335–1355.(doi:10.1023/A:1023617117780)

Whitten, Stuart, Salzman, James, Shelton, Dave and Wendy Proctor. 2003. Markets for Ecosystem Services: Applying the Concepts. Canberra, Australia: Commonwealth Scientific and Industrial Research Organization (CSIRO).

Williams, J.N., A.D. Hollander, A.T. O'Geen, L.A. Thrupp, R. Hanifin, et al 2011. Assessment of carbon in woody plants and soil across a vineyard woodland landscape. Carbon Balance and Management 6(11): (9 November 2011) London, UK: BioMed Central Ltd.

Wittenberg, R., and M.J.W. Cock. 2001. Invasive alien species:a toolkit of best prevention and management practices. Wallingford, UK: CAB International.

Wojtkowski, P.A. 2004. Landscape agroecology. New York, NY: Food Products Press.

WRI (World Resources Institute). 2005. World Resources 2005: The Wealth of the Poor – Managing Ecosystems to Fight Poverty. World Resources Institute, Washington.

WWF (WorldWide Fund for Nature). 2005. Title..494 S. J.

Yiridoe, E.K., D. Odwa, A. Atari, R. Gordon, and S. Smale. 2010. Factors influencing participation in the Nova Scotia Environmental Farm Plan Program. Land Use Policy 27(4):1097 1106. Oxford, UK: Elsevier Ltd.

Young, O.R., F. Berkhout, G.C. Gallopin, M.A. Janssen, E. Ostrom, and S. van der Leeuw. 2006. The globalization of socio ecological systems: An agenda for scientific research. Global Environmental Change 16:304 316.

Zhaoa Chuang, Liub Bing, Piaoa Shilong, Wanga Xuhui, Lobelli David B., Yao Huangj, Huanga Mengtian,Yaoa Yitong, Bassuk Simona, Ciaisl Philippe, Durandm Jean Louis, Elliottn Joshua, Ewertp Frank, Janssensr Ivan A.,Lis Tao Lint, Erda, Liua Qiang, Martreu Pierre, Müllerv Christoph, Penga Shushi, Peñuelasw Josep, Ruaney Alex C.,Wallachz Daniel, Wangg Tao, Wua Donghai, Liua Zhuo, Zhub Yan, Zhua Zaichun, and Assengf Senthold,Temperature increase reduces global yields of major crops in four independent estimates, Edited by B. L. Turner, Arizona State University, Tempe, AZ, and approved July 10, 2017 (received for review January 31, 2017)

Websites :

URL: http://journals.cambridge.org/action/displayJournal?jid=RAF URL: http://onlinelibrary.wiley.com/journal/10.1111/(ISSN)1467 2979URL: http://www.sciencedirect.com/science/journal/03014797

URL: http://www.sciencedirect.com/science/article/pii/S0921800913003595 URL: http://www.swcs.org http://www.biodiv.org/decisions/default.asp?decZVI/9.

URL: http://www.sciencedirect.com/science?_ob=ArticleURL&_udi=B6VDY 4JWFH1G 1

URL: http://www.sciencedirect.com/science/journal/0308521X.

URL: http://www.forestry.umn.edu/publications/staffpapers/Staffpaper206.pdf URL: http://www.fisheries.org/afs/docs/fisheries/fisheries_3405.pdf

URL: http://onlinelibrary.wiley.com/journal/10.1002/(ISSN)1932 1031

URL: http://www.informaworld.com/smpp/content~db=all~content=a928331368~frm=titlelink URL: http://pdfserve.informaworld.com/216710_901698885.pdf

URL: http://www.joe.org/joe/2012october/iw5.php

URL: http://www.nature.com/ismej/journal/v6/n12/full/ismej201284a.html URL: http://www.informaworld.com/smpp/title~content=t713666717~db=all www.diversitas international.org/docs/Inter.%20Diversitas.pdf.

URL:http://www.cabi.org/CABeBooks/default.aspx?site=107&page=45&LoadModule=PDFHier&BookID=840

URL: http://www.tandfonline.com/loi/wcim20

URL: http://www.sciencedirect.com/science/journal/0308521X

URL: http://www.sciencedirect.com/science/article/pii/S002216941100878XURL: http://www.esajournals.org/doi/full/10.1890/110230

URL:http://www.springerlink.com/content/05186h401164j685/?p=e1ad747f25964a318dc1198904c1cbad&pi=3

URL: http://jel.oxfordjournals.org/

URL:http://www.bioone.org/doi/abs/10.2993/0278 0771 29.2.339

URL:http://www.scirp.org/Journal/PaperDownload.aspx?FileName=JWARP20101100001_61841300.pdf&paperID=3184

URL: http://www.sfaa.net/ho/

URL:http://www.odi.org.uk/sites/odi.org.uk/files/odi assets/publications opinion files/8190.pdf

URL: http://www.agdevjournal.com/attachments/112_JAFSCD_McIntyre_Column_08 10.pdf

URL: http://www.eianz.org/

URL: http://www.ecologyandsociety.org/vol18/iss4/art20/ http://rstb.royalsocietypublishing.org/ http://www.rainforest alliance.org/cspa.cfm. http://dx.doi.org/10.1111/1462 2920.12137.

http://www.consecol.org/vol3/iss2/art5/.

http://www.cocoafederation.com/issues/stcp/.

URL: http://onlinelibrary.wiley.com/doi/10.1002/9780470960257.ch3/summary.

http://www.eologyandsociety.org/vol9iss2/art3/.

http://www.ecologandsocity.org/vol19/iss2/art5/.

http://www.ecologyandsociety.org/vol11iss1/art13/.

https://en.wikipedia.org/wiki/West_Bengal

URL:http://www.sciencedirect.com/science?_ob=ArticleURL&_udi=B6V91 4YYWRJC 1

URL: http://www.cbmjournal.com/content/pdf/1750 0680 6 11.pdfwww.wwfchina.org/wetlands

URL: http://www.sciencedirect.com/science/journal/02648377

www.thehindu.com

ANNEXURE I

DEPARTMENT OF AGRICULTURAL EXTENSION

Bidhan Chandra Krishi Viswavidyalaya, Nadia, West Bengal

Personal Interview Schedule for Data CollectionConducted by Moumita Gupta

Name of the village...

G.P.. Block........................... District..............................

Name..

1. **Age....**Chronological Age of respondent measured in year and month
2. **Education....**Year of schooling
3. **Functional Education....(** No. of training and exposure to non formal educational courses and duration)
4. **Size of family members**.... Count of family members.
5. **Size of holding**.... Total area of land owned by the respondent divided by number of family members
6. **Number of fragment**...Arithmetic count of pieces of the land(fragments) under the ownership of family.
7. **Average size of fragment**... Measured area of each of the fragments/ no. of fragments
8. **Mean distance between two fragments**...Average distance lying between 2 fragments measured on average perceived distances
9. **Size of homestead land**... Total area of land surrounding the dwelling hose owned by a family / family size

10. **Yield of Crop....**Kg/ha (Rice+vegetables+pulses)
11. **Marketed surplus...** Quantity of yield proportionate to total yield as already disposed off to the market in Kg
12. **Land under irrigation**....Size of holding in ha under irrigation
13. **Rain fed area**....Size of holding under rainfed condition
14. **Cropping intensity....**(GCA/NCA)100
15. **Total amount of seed used**....Kg/ha
16. **Total hours of irrigation given**.... Total number of hours for 3 crop in a season
17. **Total amount of chemical fertilizer applied**....In total (rice + vegetables + pulses) in kg/ha
18. **Count of livestock**....Total number of cattle, poultry and goat divided by number of family member
19. **On farm Income**...... The income expressed in monetary value generated from unit area of farm.(Rs/month)
20. **Communication Variable....**

Tv/radio	Newspaper or any print media	Dealerss or peer farmer	Experts from an Institution	Calculation
A days in a week	B days in a week	C days in a week	D days in a week	A*12+B*12+C*12+D*12

21. **Consumption of coal, fuel wood, fire wood etc....**Average in kg
22. **Production of cowdung/poultry liter /goat/biogas slurry** Average in kg
23. **Application of organic manure....**Kg/ha
24. **Rice stubble height left in the Field**....in inch
25. **Crop residue left in the field....**Raw weight of Rice straw, vegetable and pulse crop part which is not harvested in kg/ha
26. **Soil conservation level**....Percentage of area under Mulching, land shaping, contour bunds, paira cropping divided by area under cultivation.
27. **Crop Biodiversity conservation....**through Simpson index
28. **Climate resilient technology**....

    Score 3 = cultivating pea or sesamum instead of boro rice. Score 2= Changes in the water management practices.

    Score 1= Not cultivating crop which are vulnerable prevailing climate and pest and disease infestation.
29. **Ecosystem output**...Yield + raw weight of rice straw and stubble+ raw weight of parts of cauliflower and lentil which are not harvested in kg/ha

ANNEXURE II

SCHEDULE FOR ITEM ANALYSIS AND EXPERTS COMMENTS

Soil Conservation item	MR	R	SR	LR
1. Soil conservation helps protection of fertility of soil , productivity of crop thereafter.				
2. Soil Conservation also helps protection of microorganisms as well.				
3. We have enough of traditional knowledge on soil conservation, but needs proper technical support.				
4. Mechanized farming is responsible for soil erosion and compaction.				
5. Application of indiscriminate or huge amount of agrichemicals is responsible for degeneration of productive behavior of soil.				
6. For conserving soil in a unique way, every farmer should be rewarded/ recognized.				

Water Conservation item	MR	R	SR	LR
1. Water conservation helps life saving irrigation during critical drought period.				
2. Destruction of natural water bodies amounts to loss of plant and fish biodiversity.				
3. Water conservation knowledge must focus on judicious use of ground water, more on rainwater harvesting.				
4. Water, Soil and biodiversity conservation go hand in hand.				
5. Recycling of waste water is of prime importance for sustainable agriculture.				

Biodiversity Conservation item	MR	R	SR	LR
1. Crop biodiversity conservation helps stability in production, better risk management and effective use of natural resources like land, water etc.				
2. Crop biodiversity conservation also helps protection of target crops from disease and pest attack.				

3. We have enough of traditional knowledge on crop biodiversity conservation, but needs proper technical support.				
4. Destruction of crop biodiversity amounts to loss of productivity, ecological functions as well.				
5. Indiscriminate use or application huge amount of agrichemicals is responsible for degeneration of productive behavior of natural resources.				
6. Water, Soil and biodiversity conservation go hand in hand.				

Soil Stewardship item	**MR**	**R**	**SR**	**LR**
1. When soil is made productive, it clearly contributes to better farm economy.				
2. For sustainable agriculture, productive function of soil needs resilience.				
3. Soil Stewardship is close to livelihood and economy wherein conservation is closure to natural resource and its ecology.				
4. Reclamation and restoration of soil properties should be counted in terms of costs and risk implication, exactly here the stewardship function has to contribute more.				
5. Recyclability and Restoration of soil properties should be counted in terms of costs and risk implications, exactly here the stewardship function has to contribute more.				

Water stewardship item	**MR**	**R**	**SR**	**LR**
1. Acknowledge indigenous knowledge, customary laws and informal institution in water management.				
2. Conjunctive use of water increases both economic and ecological proficiency.				
3. Water Stewardship helps attaining better ecological productivity, economic growth and resilience.				
4. A well planned water use in Agriculture can generate good score of livelihood as well as wage security.				
5. Water should not be perceived as a "free gift" of nature but a resource to conserve, recycle for livelihood and sustainable growth.				

Biodiversity Stewardship item	**MR**	**R**	**SR**	**LR**
1. Crop biodiversity per unit area reduces risks and enhances profitability.				
2. Biodiversity should be conserved because it fosters ecological resilience as well as productivity.				
3. A crop mix has got certain benefits over monolithic agricultural because it integrates benefits and countervails negativity.				
4. The more the biodiversity per unit area, the higher in the economic and ecological functions trapped per unit area.				
5. Wider genetic base, as defended in higher crop biodiversity, opens up the scope for better economy and growth.				

Mr....Most Relevant (4)

R....Relevant (3)

Sr...Somehow Relevant (2)Lr.... Least Relevant (1)

INDEX

I

L

M

N

O

P

R

S

T

W

Z

9 789354 615412

Printed at Repro India Ltd.